Planning, Building and Designing with Water

Water-scapes Innova-tion. 水敏性创新设计

Herbert Dreiseitl & Dieter Grau
赫伯特·德赖赛特尔和迪特尔·格劳 编著

辽宁科学技术出版社

除书中特别提到的文字和图片版权外，所有材料的版权均属于德国戴水道设计公司（Atelier Dreiseitl）所有。对于全部或部分材料的翻译、再版、转载，图片资料的重复使用、传播、制作缩微胶片以及以其他任何方式进行的复制、数据存储，版权所有方保留所有权利。对于本书内容任何形式的使用，必须获得版权拥有者的许可。

All material's copyright will stay with Atelier Dreiseitl GmbH except those mentioned credits for text and images. All rights are reserved, whether the whole or part of the material is concerned, specifically the rights of translation, reprinting, re-use of illustrations, recitation, broadcasting, reproduction on microfilms or in other ways, and storage in data banks. For any kind of use, permission of the copyright owner must be obtained.

编委会：迪特尔·格劳、格哈德·豪博、孙峥、里奥纳德·安、托比亚斯·鲍尔
策　划：高枫、杰西卡·里德、茱莉亚·德赖赛特尔、布朗温·谭
网　址：www.dreiseitl.com

Editorial Board: Dieter Grau, Gerhard Hauber, Zheng Sun, Leonard Ng, Tobias Baur
Curators: Feng Gao, Jessica Read, Julia Dreiseitl, Browyn Tan
Website: www.dreiseitl.com

图书在版编目（CIP）数据

水敏性创新设计 /（德）德赖赛特尔，（德）格劳编著；高枫译. -- 沈阳：辽宁科学技术出版社，2014.3
ISBN 978-7-5381-8461-7

Ⅰ.①水… Ⅱ.①德… ②格… ③高… Ⅲ.①城市—理水（园林）—景观设计 Ⅳ.①TU986.4

中国版本图书馆CIP数据核字(2014)第018651号

出版发行：辽宁科学技术出版社
　　　　　（地址：沈阳市和平区十一纬路29号　邮编：110003）
印 刷 者：利丰雅高印刷（深圳）有限公司
经 销 者：各地新华书店
幅面尺寸：225mm×285mm
印　　张：13.5
插　　页：4
字　　数：50千字
印　　数：1～1200
出版时间：2014年3月第1版
印刷时间：2014年3月第1次印刷
责任编辑：陈慈良　宋丹丹
封面设计：杨春玲
版式设计：杨春玲
责任校对：周　文
书　　号：ISBN 978-7-5381-8461-7
定　　价：258.00元

联系电话：024-23284360
邮购热线：024-23284502
E-mail：lnkjc@126.com
http://www.lnkj.com.cn
本书网址：www.lnkj.cn/uri.sh/8461

Planning, Building and Designing with Water

Waterscapes Innovation. 水敏性创新设计

Herbert Dreiseitl & Dieter Grau
赫伯特·德赖赛特尔和迪特尔·格劳 编著

辽宁科学技术出版社

Waterscapes Innovation | 004

Contents

Masterplanning

- 006 Preface *by Dieter Grau*
- 010 Introduction *by Herbert Dreiseitl*
- 014 Water – the Key Resource for China's Future *by Dr. Eduard Kögel*
- 024 Central Watershed Masterplan, "Active Beautiful, Clean", Singapore
- 032 Tianjin Eco City
- 036 Tianjin Hexi CBD Landscape Master Planning
- 042 Development Masterplan, Pingdingshan, Henan, China
- 046 Restoration of Emscher River, Ruhr Valley, Germany
- 048 Offenbach Harbour, Offenbach, Germany

Ecocity District

- 052 Eco Quarters – How to Build Truly Sustainable? *by Gerhard Hauber*
- 056 Ecological Infrastructure & Stormwater Management in Tianjin Cultural Park
- 064 Water System Potsdamer Plaza, Berlin, Germany
- 070 Transformation of a City District, Taipei, Taiwan
- 078 JTC Clean Tech Park, Singapore
- 082 Kallang Riverside, Singapore

River Restoration

- 088 New Design Culture for Urban River Spaces *by Antje Stokman*
- 092 Bishan-Ang Mo Kio Park and Kallang River, Singapore
- 100 Daylighting of Alna River, Holalloka, Oslo, Norway
- 104 Restoration of River Volme, Hagen, Germany
- 108 Rochor Canal, Singapore
- 112 Feng Chan River Restoration, Zhangjiawo New Town, Tianjin, China

Urban Space

- 116 Cities Moving toward Sustainability *by Khoo Teng Chye*
- 120 Zollhallen Plaza, Freiburg, Germany
- 124 Water Traces, Hannoversch Münden, Germany
- 128 Heiner-Metzger-Plaza, Neu-Ulm, Germany
- 132 Water Management in McLaren Technology Center, London, UK
- 136 Mixed-Lifestyle Development, Changchun Vanke, China
- 140 Mittelstrasse, Gevelsberg, Germany

Urban Park

- 144 Integrated Stormwater Management Strategy for Ecological City *by Wu Che, Yang Zhao, Pan Yan*
- 150 Tanner Springs Park, Portland, USA
- 154 Queens Botanical Garden, New York, USA
- 158 Scharnhauser Park, Ostfildern, Germany
- 162 Green Roof, Chicago City Hall, USA
- 164 Maybach Center of Excellence, Stuttgart, Germany

Residential Projects

- 168 Think Global Act Local *by Prof. Wolfgang F. Geiger*
- 174 Solar City, Linz, Austria
- 178 Zhangjiawo Neighbourhood Community, Tianjin
- 184 A Climate Adaptation Community, Arkadien Winnenden, Germany
- 190 Guorui International Investment Square, Beijing, China

Art and Fountains

- 194 Art as an Icebreaker *by Herbert Dreiseitl*
- 196 Water playground BUGA, Koblenz, Germany
- 198 Taipei Public Art Project – Water-art Red Snake
- 200 Art Wall, Tanner Springs Park, Portland, USA
- 202 Central Fountain, Ritz Carlton, Tianjin Taiandao, China
- 204 Light and Water Installation, Breiteweg
- 206 Red Swings, Venlo, Netherlands
- 208 Water Curtain Fountain, Heiner-Metzger-Plaza, Neu-Ulm, Germany
- 210 Illustration Credits
- 214 The Authors

目录

	008	前言 迪特尔·格劳
	012	简介 赫伯特·德赖赛特尔
	020	水——关乎中国未来的关键资源／爱德华·科格尔博士
规划设计	024	新加坡中央地区水环境规划——"活力、美观、洁净"水敏城市设计导则
	032	天津生态城景观规划
	036	天津河西中央商务区整体景观规划
	042	河南平顶山生态景观与水文发展规划
	046	德国埃姆舍河流域修复规划
	048	德国奥芬巴赫港口区域规划
生态区域设计	054	生态区域——如何建立真正的可持续发展？／格哈德·豪博
	056	天津文化中心生态水环境基础设施
	064	德国柏林波茨坦广场水系设计
	070	台北城市街区的转变
	078	新加坡JTC清洁科技园
	082	新加坡加冷河滨区域规划
河流修复	090	城市河流空间的崭新设计理念／安茱·斯托克曼
	092	新加坡碧山宏茂桥公园和加冷河修复
	100	挪威奥斯陆海伦拉卡，埃纳河的日光
	104	德国哈根市沃尔姆河修复
	108	新加坡梧槽运河修复
	112	中国天津张家窝新镇，丰产河修复
城市空间	118	迈步走向可持续化城市／邱登才
	120	德国弗莱堡市扎哈伦广场
	124	德国汉恩蒙登"水之印"广场
	128	德国新乌尔姆，海纳–梅茨格广场
	132	英国伦敦麦克拉伦技术中心水管理项目
	136	中国长春万科综合生活区开发项目
	140	德国格勒斯伯格市中央街设计
城市公园	148	生态城市雨水综合管理策略／车伍、赵杨、闫攀
	150	美国波特兰坦纳斯普林斯公园
	154	美国纽约皇后植物园
	158	德国奥斯特菲尔登沙恩豪瑟公园
	162	美国芝加哥市政厅屋顶花园
	164	德国斯图加特迈巴赫卓越中心
住宅项目	172	放眼全球·本土行动／沃夫冈·F·盖格教授
	174	奥地利林茨太阳城
	178	天津张家窝新镇社会山小区
	184	德国阿卡迪亚温嫩登气候适应型社区
	190	中国北京国锐国际投资广场
艺术水景设计	195	让艺术成为设计的破冰者／赫伯特·德赖赛特尔
	196	德国科布伦茨BUGA嬉水游乐场
	198	中国台北公共艺术设计——红蛇水景艺术装置
	200	美国波特兰纳斯普林斯公园艺术墙
	202	中国天津泰安道丽思卡尔顿酒店中央水景
	204	布莱特威戈"光与水"的装置
	206	荷兰芬洛红色摇摆水景装置
	208	德国新乌尔姆，海纳–梅茨格广场，水幕喷泉
	210	图片版权
	215	作者简介

Preface
by Dieter Grau

Working with urban water presents one of the most fundamental and rewarding challenges in our cities today.

After realizing Asia is heavily in need of flood protection and improvement of water quality and urban quality, we feel delighted to present our ideas for those issues in this book. After the three editions of Waterscapes, we made a conscious decision to publish our most current work now in an English/Chinese bilingual version for Asia. As the urbanisation process has reached an enormous speed, the consideration of clean water, blue sky and quality of life for people in dense cities has reached utmost importance. This has to be seen also in the light of a harsh competition between cities and their abilities to attract investment and talent.

Creating a better systemic approach to resources in cities, where the local people and local authorities are integral part of the solution is one of the key directions, in simple words this could be recalled as a process "vision creation" instead of "problem solving". Low-tech urban innovations are necessary to enhance the performance of cities in the future but consider the underdeveloped knowledge of those systems maintenance with non educated people.

We have focused in this book on project samples which show a sustainable approach to new waterscapes for cities and even extending it to a regional scale. The core approach of all the work shown is the deep integration between city planning and infrastructure. But an ultimate component of liveable cities is the human scale quality, where water can play a fascination role and can provide beauty and aesthetical qualities.

Some samples address key issues of flood management and water quality, while seeking to create the greatest possible synergy with the urban environment. This new generation of urban waterscapes, or blue-green infrastructures addresses essential city services such as mobility, recreation, safety and biodiversity, creating a strategic and feasible approach to ensure long-term resilience and economic buoyancy. The networking of city public space as interactive, ecological infrastructure makes our cities more beautiful, functional, safe and comfortable. This means that parks, plazas, streetscapes and rivers are blue-green infrastructure. Every investment is rewarded by the social-economic synergies of this interdisciplinary approach.

Take, for example, Singapore. The Kallang River- Bishan-Ang Mo Kio Park project is a new vision for blue-green city infrastructure which addresses the dual needs of water supply and flood protection while creating spaces for people and nature in the city. Tianjin Cultural Park is the city's most significant urban redevelopment project which manages stormwater for microclimate improvement and flood protection, while creating a high quality civic stage for Tianjin City's new cultural center. Offenbach Harbour represents a pragmatic vision for how climate adaptation can be achieved in urban context creating attractive neighbourhoods through integrated engineering.

Thanks to the great support of all colleagues in the Atelier studios, special thanks to Dr. Gao Feng and her continued work on the translation and coordination. Many thanks also to all our friends and colleagues who took the effort and contribute with great passion. Special thanks to Dr. Eduard Kögel, Gerhard Hauber, Mr. Khoo Teng Chye, Prof. Wolfgang F. Geiger, Prof. Antje Stockmann, and Prof. Che Wu for their essays as an essential part of the book.

前言
迪特尔·格劳

应对城市水环境问题成为了当今城市之中最根本、最具挑战意义的课题之一。在了解到亚洲国家在抵御城市洪水侵袭、改善水质和提升城市品位的强烈需求之时，非常荣幸能够在这本书中介绍我们在面对这些问题方面的观点。在出版此书的前三个版本之后，我们决定在这本面向亚洲国家的中英双语的书籍中更新、发表最新的代表作品。在亚洲国家城市化进程正值高速行进之时，对于洁净的水体、蔚蓝的天空和改善人口密集城市之中生活质量的考虑已经极为重要。它们同样也被视为城市之间在吸引投资和引进人才方面激烈竞争的关键要素。

创造一个城市之中更加完备的资源系统解决方案，将地方居民和政府机构整合作为解决方案的组成部分，成为了解决方案的关键步骤。简单而言，这并非单一的"解决问题"的过程，而可以被重新定义为一个"视觉创造"的过程。当前较低技术含量的城市创新设计尽管在提高未来城市机能方面发挥作用，但仍需意识到在这些系统的设计和维护方面从业者知识和经验的欠缺。

在这本书中，我们专注于项目案例，展示了以可持续的方式设计城市以及城市区域范围的水敏性景观。而这些项目的核心则归结为城市规划与基础设施的深度整合。不过，宜居型城市的最基本内容仍为人性化尺度的设计质量。在那样的空间之中，水敏性设计可以成为一种极具吸引力的元素，提供美观、亮丽的城市环境。

在一些案例之中，明确提出了完善洪水管理、提升城市水质，并同时寻求创造水资源与城市环境协同作用的最大可能等关键性问题。这样一种新生代的城市水敏或蓝绿基础设施设计，满足了如流通、休闲娱乐、安全和生物多样性等方面的基本城市服务，能够创建以一种战略性、可行的方式保障长效城市弹性适应能力和经济增长浮力。互动、具备生态基础设施特征的城市公共空间网络，能够让我们的城市更美丽、功能齐备、安全和舒适。它也意味着公园、广场、街景和河流将共同组建成为城市之中的蓝绿基础设施。每一项投资都会从这种跨学科的经济、社会协同作用之中获益。

新加坡，加冷河碧山宏茂桥公园项目是一个全新的城市蓝绿基础设施的典型案例，满足了水资源供应和防洪保护的双重需求，同时为城市之中的人们创造了与自然接触的空间。天津文化中心作为天津市近十年来最重要的城区重建项目，管理雨洪、提升区域小气候、进行防洪保护，并为天津市新的文化中心区域创造了一个高品质的市政平台。德国奥芬巴赫港口区域的规划设计则展示了如何在城市背景下实现气候适应性设计，并通过设计与工程的整合、创造富有吸引力的邻里社区。

感谢德国戴水道设计公司全体员工的大力支持，特别感谢高枫博士和她一直以来的翻译和协调工作。同时十分感谢我们所有的朋友和同事们的努力和富有激情的工作。特别感谢为此书写作主旨文章的爱德华·科格尔博士，格哈德·豪博先生，邱登才先生，沃夫冈·F·盖格教授，安荣·斯托克曼教授，车伍教授。

Introduction
by Herbert Dreiseitl

Waterscapes have always been the driving component for structures in natural and urban environments. Without water there is simply no life. Never bound to strict limitations water fulfills in the long term the innovative character of transition, change, and development not only in the natural environment but also within urban settlements.

We have been active with integrated water-projects in cities all around the world for more than 30 years, combining contextual design with sustainable engineering. This was not a given standard, and waterscapes in cities were seen more as a challenge and danger instead of an opportunity. To drain and to get rid of water as fast as possible was a common practice. In the beginning of our work engineers were making jokes about our ambition to find better ways of environmentally responsive water management. Today this is a mega-trend and we can see this crucial topic is receiving more and more awareness in media, politics, and in the professional world, one we foresee increasing even more as an urgent topic in the future.

Natural water systems always combine functionality with a high aesthetic performance. There is a sense of logical balance, of "common sense" in these natural hydraulic systems that can create adaptable and flexible opportunities in dynamic landscapes where many vibrant life forms can exist and even depend upon the environment. It is a creative beauty that takes spaces and performs at the appropriate time, a motion we have often lost within industrialization, our artificial technology, and rising urbanization.

To bring us closer to this natural prototype it always comes back to a fundamental question: Can we create living systems that filter, clean and regulate water, balance temperature, produce good air, save natural resources, increase biodiversity - and what are the basic principles and processes to achieve this while integrating in space and time?

A significant amount of research and development today has gone into engineering. Not only are mathematic hydraulic models compiled or flood risk analysis and protection created but strategies for traffic control, energy efficiency, and other urban design and planning tools. Often we know about these methods, engineering tools, and instruments but we do not know how to implement this in the urban fabric. There is a competition for space usage especially in urbanized areas, and this conflict is dramatically rising in turn with increasing urban density. Traffic, industry, shopping, housing, parks, recreation, and leisure activities are growing - but in the near future we will simply run out of space. Each of the aforementioned functions and services demanded its own rights and space needs, and tended to be located and treated separately without any relationship to context. With this attitude many cities were created and look today like old, inflexible, broken machines that need permanent reparation and demand expensive maintenance. Even with the best intentions in mind, this service draining our limited resources, deteriorating the environment, and destructing sustainable urbanization and livable cities.

Can we afford this in the future?

The key seems to be found within the urban performance and how to best integrate adaptable, flexible infrastructure. Cities will need to become more multifunctional with its allocation of shared spaces. Simultaneously cities will need to become more kinetic, utilizing spaces at different times with varying, different functions. Like space is needed for water after a rain event, the dynamic occurrence of multiple mobility at rush-hours and gatherings of immense crowds of people for a special cultural event is also in need of space. We have to invent cities that can handle these processes. The key lies then therefore in not just technology but meaningful and resilient solutions of the urban form by good governance, new forms of organization, and integrated aesthetically attractive design enhancing quality of life standards.

Combining high forms of aesthetics and efficient technical performance works best within the local cultural context. We have seen too many glittering, slick masterplans with a strong top down attitude that have in the end proven ineffective and have failed. Since our establishment in 1981 we have been constantly searching for opportunities to integrate participation and processes, to bring the local residents and stakeholders on board right from the beginning. Our projects were successful when we were able to find the right mix of a bottom-up-and-top-down approach, with centralized and decentralized organizational structures. Together we pushed the envelope and created, as pilot projects, many resilient shared spaces in cities. This resulted in a win-win situation for the local people and the natural environment.

Innovative Waterscapes influence modern lifestyles, increase better health conditions, and create more beauty in cities. Instead of the hard, inflexible, and old machines of the past, Cities of the Future can become resilient dynamic organisms supporting our modern urban life and society.

简介
赫伯特·德赖赛特尔

水敏性设计一直作为自然和城市环境结构之中的驱动组件。没有水就没有生命。水从不受限制，长期以来存在于自然以及城市居住环境之中，不停歇地转换、变化和发展，实践着自我的更新。

三十多年来，我们一直活跃在世界各地城市综合水务项目的设计活动之中：将可持续的工程实践与广阔背景下的概念设计结合。这里不存在一个既定的标准，现存城市之中的水敏性设计更多的是被看做一种挑战和危险，而非机遇。排放、尽快去除多余水体是一种常见的做法。在我们工作的初始阶段，工程师曾经对于我们寻找更好的环境适应型水管理方式的努力付之一笑。而今天，这已经成为了一种大趋势，我们可以看到这一至关重要的话题正在受到媒体、政界以及专业领域越来越多的关注，我们完全可以预见到关于水敏性设计的话题在未来会变得愈发的紧迫和关键。

自然水系总是能够将功能性与极高的审美表现结合起来。这里有一种逻辑上的平衡，一种自然水利系统之中的"常识"，它能够创建动态景观之中充满适应性和灵活性的机会，充满活力的生命形式便可能存活并与环境相互依存。它是一种创造性的美，能够在适当的时间出现并发挥作用；它也是一种在我们的工业化、人工技术迅猛发展以及不断浓重的城市化之中常常被丢失的一种态度。

为了使我们更接近这一自然的原型，又回到了一个基本问题：我们是否能够创造具备过滤、净化以及调节水量、平衡温度、制造清洁空气、节约自然资源、提高生物多样性的生命系统？而整合空间和时间，达成这些目标的基本原则和流程是什么？

如今的大量研究和发展已经进入工程领域，不仅在数学水力模型建造或是洪水风险分析和保护方面，而且在交通控制、能源效率以及城市设计和规划工具应用策略上都有所运用。通常情况下，我们也许仅仅了解这些方法、工程工具和仪器设备，但却并不知道如何将其在城市结构之中加以实践。在空间使用特别是城市地区空间使用方面存在着竞争，且随着不断增加的城市密度，这种竞争冲突变得愈发剧烈。交通、工业、商业、住房、公园、娱乐和休闲活动不断增加，在不久的将来，我们可用的空间将变得更加紧张。上述的每一项功能和服务都有其自身的权利和空间需求，并且常常孤立分布、彼此与周围环境之间无任何关联。在这样的态度指导之下，许多城市被建造起来，在今天看来就如老旧、机械、破碎的机器，需要永久的修缮和昂贵的维护费用。即使拥有最美好的意愿，这种服务也在耗尽我们有限的资源、恶化环境、破坏可持续的城市化和宜居城市发展。

未来，我们是否能够担负？

问题的关键很大程度上在于城市的表现，以及如何最好地整合适应性强、灵活的基础设施。城市需要变得更加多功能，享有更多的公共空间。同时，城市需要变得更具动感，在不同的时间内利用空间发挥多样变化的功能。正如降雨之后所需要的雨水排放、城市峰值时刻发生的紧急事件以及特定的文化性集会时的人群快速聚集活动等，都需要城市里的缓冲空间。我们必须创造出能够应对这些状况的城市空间结构。因此，问题的关键不仅依赖技术层面，而且包含对于城市形式的富有意义的弹性解决方案。通过良好的管理，新型的组织形式，综合、美观、且具吸引力的设计，来提高生活质量标准。

综合高品质美学形态和高效的技术性能的作品最能与地方的文化背景完美结合。而相反，我们已经能够清楚地看到，太多被光环笼罩、目标宏伟远大、但却华而不实的规划项目最终被证明无效而失败。自德国戴水道设计公司于1981年成立以来，我们持续不断地寻找机会注重参与和过程，从初始阶段即寻求带给当地居民和项目业主们广泛的权益。我们总是能够找到由下至上以及自上而下的正确组合方式，运用集中和分散的组织结构，保证项目成功。我们以此为基准，创建众多的试点项目，组成了城市之中的弹性共享空间，如此便形成了当地居民和自然环境之间双赢的局面。

《水敏性创新设计》一书之中所介绍的设计准则和项目案例能够影响现代生活方式、提高环境卫生状况、创造城市之中更多的美景。取代坚硬、毫无活力、老式的城市机器，未来之城可以成为富有弹性的动态有机体，以支撑我们现代化的城市生活和社会发展需求。

Water – the Key Resource for China's Future
by Dr. Eduard Kögel

"Since 1991, 31 million hectares of drought-affected farmland are irrigated annually on average, saving 39.41 million tons of grain, and enabling 24.36 million people and 19.08 million livestock access to drinking water each year." This statement is taken from the website of the Ministry of Water Resources of the People's Republic of China and reveals one of the big water related problems the country is facing. Water consumption for agriculture is between sixty and seventy per cent of total consumption, whereas industry uses around twenty-five per cent, leaving only twelve per cent for domestic use. The national policy for food self-sufficiency in a rapid urbanization has significant effects on irrigation-based agricultural production. More than forty per cent of the population lives in the North China Plain with only eight per cent of the national water at their disposal.

The uneven distribution of water between North and South China with a still-expanding population, poor water related infrastructure and management, pollution of surface water and depletion of groundwater as well as regular floods and droughts, result in growing regional conflicts over water-allocation. Experts believe that the rapid economic development of China will be substantially effected by these challenges. Many of the problems have to be tackled on an administrative level, but others only can be changed by individual awareness and modification of behavior. Both aspects must be seriously considered in the future.

Traditional Aesthetic and Technical Concepts

In ancient times the treatment of water related issues was already a topic in intellectual discourse, and was used in different fields of artistic expression and religious practice. The heroic hydraulic engineer Li Bing and his son Li Er Lang built the Dujiangyan Irrigation System in Sichuan almost 2,300 years ago. Their weir system was ingenious and, over time, was further developed into a distribution network of water in the basin of Sichuan, which gave rise to a rich agricultural production. This ancient system is still in use today. No wonder that the two engineers were idealized and worshipped in temples dedicated to their name. Hydraulic engineering and irrigation became a key issue for the development of a complex agricultural society in ancient China.

Early on, the rice-growing regions of Asia developed communal management systems for the distribution of water. In the case of China and India, the historian Karl August Wittfogel spoke of "hydraulic empires". The natural phenomenon of the monsoon with its seasonal rainfall and the reliance on continual water supply for agricultural production made it necessary to base the management on a powerful bureaucracy. The centralized water administration was the structure for what Wittfogel called "Oriental Despotism".

In the 1930s, in his work "Climate and Culture", the Japanese philosopher Tetsuro Watsuji pointed out the interconnectedness between climate, environment and human culture, in order to understand the basis of regional Asian societies. In the case of central China he spoke of the "monsoon society", mainly depending on the regular seasonal rainfall with its specific need for management of water, which had the consequence of a less individual, but more communal society.

Throughout history, the danger of floods and droughts was always present and guided the general concept of water management, not only in agriculture but also for the layout of human settlements. The awareness of water as part of the manmade environment is preserved in the traditional aesthetic concept of Chinese culture. The ideal landscape is reflected in traditional garden design, with a harmonious relationship between rocks, plants, pavilions and water.

Today the balance and harmony is ousted by technical feasibility and maximizing agricultural production in a fast growing economy. After the foundation of the People's Republic in 1949, water management became a key task for the state. The building of dams and irrigation systems, river regulation and monumental transfer projects from water-rich regions to the arid north of the country, was a political issue right from the start. Technical feasibility became the guiding principle for the experts within the political system. Hu Jintao, General Secretary of the Communist Party and President of the People's Republic of China, is a trained hydraulic engineer.

Natural Conditions and Conflicts

The huge surface area of the People's Republic of China is drained into the sea by only a few river systems. By far the biggest is the Yangtze River system in central China, which drains one-fifth of the country's surface and carries almost half of its surface water. Its source lies in the Tibetan mountains, from where the river runs 6,300 kilometers before emptying into the sea near Shanghai. Half of the country's agricultural production comes from the drainage area of the Yangtze River.

The second longest river, the 5,400 kilometer long Yellow River, empties into the Bohai Sea. In the last hundred years it has caused many deadly floods with several million victims. Currently, the lower reaches and the river mouth in Shandong Province fall dry half the year, with the consequence of further lowering the ground water level of the region. From its source in the Tibetan Mountains, it winds through nine provinces, including Inner Mongolia, Shanxi and Shaanxi, where massive erosion tints the water yellow. The large amount of sediment causes problems in reservoirs and in hydroelectric power plants. In the lower regions, where the silt deposits, an elevated riverbed exacerbates the permanent danger of flooding.

In the southern border lands conflicts arise with neighboring countries like India, Bangladesh, Burma, Thailand, Laos and Vietnam surrounding water withdrawal in the upper reaches of the Mekong, Brahmaputra and Red Rivers, which all originate in Tibet or Yunnan. International negotiations about the amount of water used by China could not yet solve the problem to the satisfaction of all affected nations. With canal projects under discussion to divert the water from the named rivers towards China's arid regions, the conflicts could become more serious in the future.

A further international conflict looms with countries like Tajikistan, Kazakhstan and Kyrgyzstan about the water from Tienshan and Pamir Mountains in central Asia. In this region China needs more water for the development in the arid land of its western province Xinjiang.

Droughts and Floods

The climatic changes in recent years have caused droughts and floods in Asia with catastrophic impact on infrastructure and arable land, like 2010 in Pakistan, where a destructive flood affected twenty million people. Neighboring regions in India suffered severe drought at the same time, due to the monsoon's erratic timing. The People's Republic of China is likewise influenced by the changes in the natural climatic system. In spring 2010 severe droughts affected several Chinese provinces: Yunnan, Guizhou, Sichuan, Shaanxi and others, with serious effects in neighboring countries like Thailand and Vietnam. The droughts in these usually water-affluent regions have been called the worst in a century.

The traditionally dry northern Chinese provinces of Shandong, Shanxi, Shaanxi, Henan, Anhui and Gansu – the breadbasket of China – were hit by a severe drought in spring 2011, the worst in the region since record keeping began in 1950. Only about ten percent of the expected rainfall for the season was measured in Shandong. In June, in some of the regions, floods replaced the droughts. Even along the Yangtze River many areas were affected this spring by the worst drought in the last 50 years. But extreme rainfall downstream in the same period, threatens heavy floods.

Water Pollution and Wastewater Treatment

In 2005 a chemical plant in Jilin City, in the northeastern province of Jilin, exploded and an estimated 100 tons of toxic pollutants entered the Songhua River. The city of Harbin, with almost ten million inhabitants, depends on the Songhua River for its water supply, which was closed down for some time after the event. Other cities along the river were also affected. The long-term consequences for the inhabitants are unknown. The drinking water supply of the city of Wuxi in Jiangsu Province, with almost six million inhabitants, which relies on the water of Lake Taihu, was seriously affected by an algae bloom, caused by a combination of pollution, sinking water levels and high temperatures in summer 2007. In both cases thousands of people left the cities, to stay in areas with a safe water supply.

In addition to such major events, the everyday practice of dumping chemical wastewater unfiltered into rivers, causes major conflicts in society. Along the Yangtze Basin alone, 10,000 chemical plants release their wastewater unfiltered or uncontrolled into the rivers. It is impossible to use the water for industrial or agricultural production without prior treatment. Under such circumstances it is impossible to think about water supply for human use. The contamination of surface water and groundwater in some regions is cataclysmic. The quality measured at more than 400 flowing water-monitoring stations indicates that they are ecologically degraded. Fertilizers seriously pollute half of all natural lakes and reservoirs.

The dramatic events are followed by an outcry in the media, but after smaller incidents the unrest among citizens is also growing in many towns and regions. Each year thousands of public protests about water issues occur across the country. The vulnerability of the water supply is increasingly becoming a political problem of significance. The administration is aware of the problem and many hundred purification plants are planned or under construction. But inadequate operation and management in existing plants indicate, that a further investment in soft skills like management and negotiation is needed.

Groundwater and Water Distribution

The depletion of groundwater in northern China is a serious problem and adequate water supply to the big cities is a constant challenge. Massive urbanization and industrialization demand much more water than the groundwater and natural water cycle can provide. The surface water in northern China is naturally extremely limited and today in many cases seriously polluted. Other alternatives have to be considered. The most effective policy would be to reduce its use, especially in agriculture. The simplest method however, appears to be drilling deeper wells and taking the groundwater from non-renewable aquifers. Many of these aquifers suffer from natural arsenic contamination and affect the health of millions of people. Due to the fact that many new deep groundwater resources are tapped, the danger of poisoning is on the rise, besides the effect of further lowering of the groundwater table.

Another solution, favored by the government for years, is the transfer of water from the south to the north. Expertise in hydraulic engineering was the key in the construction of the world's longest and oldest artificial waterway, the Grand Canal, between Beijing and Hangzhou. The construction of the more than 1,700-kilometre-long canal took place between the fifth century BC and seventh century AD. Such gigantic undertakings give contemporary technicians the confidence to construct similar canals for the transfer of water. The South-North Water Transfer Project is divided into three canals – the Western Route, the Central Route and the Eastern Route. The Eastern Route depends on the Grand Canal, and water will be pumped from the Yangtze River to the northern city of Tianjin. The project is expected to start operation in 2012. The Central Route is under construction and will bring water from the Han River, a tributary of the Yangtze River, to the northern plains around Beijing. The first part is finished, and the overall project is expected to open in 2014. The Western Route is the most complicated, because neighboring countries like India, Thailand and Laos are directly affected, as already mentioned above.

Awareness and Action

A systematic approach, based on economy and efficiency, combined with information campaigns and local small-scale projects, as well as a clear and controlled water policy by the central administration, is needed to face the challenges associated with access to clean and safe water. Based on personal experience, projects such as those by Atelier Dreiseitl or the Beijing based firm Turenscape show how Landscape Architects can treat the problem in an aesthetic and mediating way. Such places, like resident gardens with clean and accessible waterways or public riverbanks to improve the green balance and the quality of live in the immediate environment, provide the direct aesthetic quality of a principal resource and will, in a playful way, hopefully, channel the responsible use and value of this natural asset within society.

Alongside the informed individual, administrative measures against inadequate management and corruption are key to the future development. Israel could act as a model for drip irrigation in the field of agriculture, for example, and Singapore could be taken as a good example of urban water management – inclusive ideal projects such as the redevelopment of Kallang River and Bishan-Ang Mo Kio Park by Atelier Dreiseitl – but in many other respects it is only our own unique experience and experimentation that will be relevant to future water projects. Besides master planning, neighborhood development and residential projects, the access of children to clean and safe water in the public space will be key for a good future, which is illustrated by the wide variety of projects in this book.

水——关乎中国未来的关键资源

爱德华·科格尔博士

"自1991年以来,受干旱影响的3100万公顷农田得到灌溉,平均每年增收粮食3941万吨,保证了2436万人次和1908万牲畜的饮用水。"该表述摘自中华人民共和国水利部网站,显示了中国正在面临的与水相关的主要问题之一。农业用水量占总消费水量的60%~70%,工业用水约占总量的25%,剩下只有12%的比例作为家庭用水。在一个快速城市化的国家中,粮食自给的国家政策对于以灌溉为基础的农业生产具有重要的影响。中国北方平原拥有超过40%的人口数量,但却仅仅拥有8%的国家水资源。

中国南北方水资源分布不均与持续增长的人口数量、水平低下的水基础设施和管理、地表水的污染、地下水的枯竭,以及定期的旱涝灾害,导致了不断增长的基于水分配问题的地区冲突。专家认为,中国经济的快速发展将会大大受制于这些方面的挑战。许多问题必须在行政级别之上予以重视,但还有一些却只能够通过个体的意识和行为的调整加以改观。未来,必须认真从这两个方面进行思考。

传统的审美和技术概念

在古代,与水治理相关的问题便已经成为技术探讨的话题,且在艺术表现和宗教实践的不同领域之中得到应用。伟大的水利工程师李冰和他的儿子李二郎在距今约2300年前修建了四川都江堰水利灌溉工程。他们的堰坝系统设计精巧,并随着时间的推移得到了进一步发展完善,形成了四川盆地的水利分配网络,为当地富饶的农业生产活动提供了条件。这一古老的系统沿用至今。正是因此,这两位工程师受到人民的尊敬,后人以他们之名建造了庙宇(二王庙)而对其加以瞻仰。水利工程和灌溉在中国古代便成为了复合农业社会发展的关键性要素。

很早以前,亚洲水稻种植区域开发了水源分布公共管理系统。以中国和印度为代表,历史学家卡尔·奥古斯特·魏特夫称其为"水利帝国"。伴随着季节性降雨的自然季风现象以及依赖持续供水的农业生产的存在,使得依托强大官僚机构为根本的管理体系成为了生产生活正常运转的必然。集中式的水管理就是魏特夫所描述的"东方专制"结构。

在20世纪30年代,在著作《气候与文化》中日本哲学家辻哲郎指出了气候、环境和人文文化之间的内在联系,以期能够解释亚洲国家的社会基础。在谈及中国中部的案例时,他说到"季风型社会"主要依赖于定期的季节性降雨来满足其具体对于水资源管理的需求,这也便是在组织形式上更多地依赖于公共社会资源而非个人之力的原因。

纵观历史,对于农业生产以及人类房屋建造布局而言,洪水和干旱的危险始终存在,并指导着水管理的整体概念。对于水作为人工环境一部分的意识,在传统的中国文化审美观念之中既已存在。理想主义的景观设计在传统园林之中得到体现,岩石、植物、亭台楼阁与水元素之间形成了和谐的关系。

今天，在经济快速增长的形势下，这种基础的平衡与和谐逐渐被愈发"先进"的技术可行性和追逐农业生产产值最大化的目标所取代。1949年中华人民共和国成立后，水资源管理即成为了国家的一项关键任务。自那时起，从水资源丰富的省份至中国北部干旱地区，水坝和灌溉系统的建造、河道整治以及巨大的水利调送项目的建设，就成为了一个政治任务被重视和执行。对于政权之内专家而言，技术的可行性也就成为了指导性原则。例如，前中共中央总书记、国家主席胡锦涛本身即为一名训练有素的水利工程师。

得天独厚的自然条件与冲突

中国国土面积上绝大部分的水体通过十分有限的几条河流系统汇集入海。到目前为止，最大的河流体系是位于中国中部的长江流域，它排放中国五分之一国土表面的水体，并运载了几乎一半的地表水。它发源于青藏高原，从那里流经6300千米，最后注入上海附近海域。中国农业产量的一半都来自于长江的冲积流域。

第二长河是5400千米长的黄河，它向东流入渤海。在过去的百年中，它已经引发了许多灾难性的洪涝灾害，数百万人因此受害。目前，其下游流域和位于山东省的河口位置在一年之中的一半时间内都处于干涸状态，导致了该地区地下水位的进一步降低。从青藏高原的源头出发，它蜿蜒穿行包括内蒙古、山西、陕西在内的九个省市，流经之地的大规模土壤流失将河水染黄，大量的泥沙沉积为水库和水力发电厂带来了麻烦。淤泥在地势较低的地区沉积下来，而河床的升高又加剧了永久性的洪涝危险。

在南部边境，围绕着起源于中国西藏及云南省的湄公河、雅鲁藏布江和红河上游流域的水源获取，中国与其邻国如印度、孟加拉国、缅甸、泰国、老挝和越南等国家之间土地纷争不断升级。关于中国消耗水量的国际性协商，仍然未能解决问题、达到所有受其影响国家的满意程度。随着正在商讨的运河项目的推进、传送上述河流水到中国的干旱地区，在未来，这种冲突可能变得更加严重。而中国与塔吉克斯坦、哈萨克斯坦和吉尔吉斯斯坦等有关国家在中亚天山和帕米尔高原一带关于水源的国际冲突同样时隐时现。在此地区，中国需要更多的水源供应以保障其西部省份——新疆干旱地区的发展需求。

干旱和洪涝灾害

近年来的气候变化已经引发了亚洲的旱涝灾害，对于基础设施和农田耕地带来灾难性的影响，比如在2010年巴基斯坦发生的破坏性洪水灾害中，受影响人数达到两千万。同时，由于季风季节的不定期性，邻近国家印度也同样遭受了严重的干旱。中国也深受自然气候系统变化的影响。2010年春季，严重的干旱影响了如云南、贵州、四川、陕西等诸多中国省份，同时对于泰国和越南等邻国也造成了严重的影响。这些在通常水资源丰富、降水量充沛的地区发生的干旱现象也被称为世纪性的灾难。

2011年春天，在传统的中国北方干旱省份和主要粮产区山东、山西、陕西、河南、安徽和甘肃，发生了该地区自1950年有记录以来最严重的旱灾。山东的实际降雨只达到此季节预期降雨量的十分之一。而在6月，一些地区的洪涝又取代了干旱。即使在长江沿岸，春季许多地区依然受到了过去50年来最严重的旱灾影响。但同一时期的下游极端降雨，又对当地造成了特大的洪水威胁。

水污染和废水处理

2005年，中国北部吉林省吉林市一家化工厂发生爆炸，据估计有100吨的有毒污染物流入松花江。因此事件，依赖松花江作为供水来源、拥有近千万居民的哈尔滨市生活供水在一段时间内被迫关闭，其他一些沿江城市生产生活用水也受到影响。而此事件对于当地居民可能造成的长期影响至今还不可知。同样，拥有近600万人口、依赖于太湖水作为饮用水来源的江苏省无锡市的饮用水情况，也因污染、水位下降以及2007年的夏季高温等因素造成的蓝藻泛滥而受到严重影响。在这两种情形之下，成千上万的人们离开城市，去寻找能够提供安全用水的地方。

除了这样的大型事件，日常倾倒未经过滤处理的化工废水、排入河流的行为，也造成了重大社会矛盾。仅沿长江流域，10,000家化工厂随意倾倒所产生的未经过滤的污水进入临近河流。这样经污染后的水源如果不经处理是不可能用于工业或农业生产的，也是不可能想象作为人类使用水源的。某些地区的地表水和地下水的污染是灾难性的，对于超过400个流水监测站的质量评估显示了水质的生态性退化。一半数量的天然湖泊和水库水质富营养化严重。

这类重大事件的发生引起了媒体的一片哗然，而在一些类似的小型事件发生之后，在许多城市和地区，人们的不满在不断增加。在中国各地，每年都有数以千计的关于水问题的公众抗议事件发生。水供应系统的脆弱性正日益成为一个重要的政治问题。政府已意识到了这个问题，几百个污水净化厂正在被规划或已在建设过程之中。但现有工厂中的不恰当操作及管理水平显示，管理和谈判的软技能培训仍然显得十分必要。

地下水资源及其分布

中国北方地区地下水的枯竭是一个严重的问题，保障大城市充足的水供应已经成为了一个持续的挑战。大规模的城市化、工业化需要比地下水和自然水循环所能够提供更多的用水量。自然情况下中国北方的地表水是极其有限的，并在现今的许多情况下受到严重污染。因此，必须考虑采用其他的替代方案。最为行之有效的政策即为减少使用，尤其是降低农业方面的用水量。而最简单的方法便是钻更深的井、从不可再生含水层提取地下水。但许多这样的含水层遭受天然的砷污染，会对数以百万计人的健康带来危害。因此，由于许多最新的深层地下水资源被开发利用，除了会带来地下水位的进一步降低，中毒的危险性也在升高。

另一种被政府多年青睐的解决方案，即是将水从南方转移到北方。水利工程方面的专业知识成为了建造起止于北京和杭州、世界上最长和最古老的人工航道——京杭大运河的关键。该1700多千米长的运河建造于公元前500年至公元700年间。如此庞大工程的成功建造为当代技术人员兴建类似的调水运河工程奠定了信心。南水北调工程分为西线、中线和东线三段。东线依托于京杭大运河，将水从长江泵出，向北方城市天津传输，在2012年开始动工。中线工程目前正在建设之中，将水从长江的支流——汉江运送至北京的北部平原，第一阶段已经实施完毕，整个项目预计于2014年竣工。而西线工程正如上文所提到的，因印度、泰国和老挝等邻国直接受其影响，情况最为复杂。

意识和行动

基于经济和效率的系统方法、结合媒体宣传、小规模的地方项目，以及中央政府制定的明确、具备管控性的政策，是应对获取清洁、安全用水挑战的必需。据我所知，德国戴水道设计公司或北京土人景观规划设计研究院所完成的项目展示了景观设计师们如何以一种审美和协调的方式处理这个问题。一些场所，比如拥有洁净、可进入式溪流水道的居住区花园或公共河岸，能够改善绿色平衡、提升居住环境质量，提供水资源的直接审美质量，以一种有趣的方式、在社区之内对这一自然资源进行负责任的引导，体现了其宝贵的价值。

除了教育讲解，针对不当的管理和腐败采取的行政治理措施是未来发展的关键。如果以色列国家可以被视为农业领域滴灌技术应用的一个典型，新加坡便可以作为城市水管理方面的一个优秀案例，它成功地展示了由德国戴水道设计公司规划设计的加冷河修复和碧山宏茂桥公园的重建项目在内的广泛优秀案例。而在其他许多方面，未来的水敏项目将会更加关注于我们自身独特的经历和实践。在总体规划、社区发展和住宅项目之外，在公共空间中儿童对于洁净、安全水源的可接近性将是实现一个美好未来的关键，这也将通过本书中类型多样的项目得到阐释。

Central Watershed Masterplan, "Active Beautiful, Clean", Singapore
新加坡中央地区水环境规划——"活力、美观、洁净"水敏城市设计导则

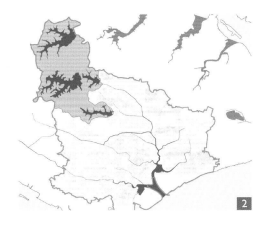

Location Singapore
Client Public Utilities Board (PUB) & National Parks Board
Partner CH2MHILL
Area 200km² (Central Catchment)
Design Time 2006
Construction Time Since 2009
项目地址 新加坡
项目委托 新加坡公共事务局和国家公园委员会
合作机构 CH2MHILL
项目面积 200平方千米（中央集水区面积）
设计时间 2006年
施工时间 2009年至今

Launched in 2006, the Central Watershed Masterplan and ABC (Active, Beautiful, Clean Waters) Urban Design Guidelines were developed by Atelier Dreiseitl as part of the city's long-term initiative to transform the country's water bodies beyond their functions of flood protection, drainage and water supply, into vibrant, new spaces for community bonding and recreation. Over 100 locations has been identified for project implementation in phases by 2030, with 20 projects already completed, bringing people closer to water. Public participation was fully integrated into the design process. Bishan-Ang Mo Kio Park and Kallang River Restoration is one of the flagship projects under this programme.

Bishan-Ang Mo Kio Park is one of Singapore's most popular parks in the heartlands of Singapore. As part of a much-needed park upgrade and plans to improve the capacity of the Kallang channel along the edge of the park, works were carried out simultaneously to transform the utilitarian concrete channel into a naturalised river, creating new spaces for the community to enjoy.

2006年开始推行的中央地区水环境总体规划和"活力、美观、洁净"城市导则由德国戴水道设计公司设计。作为城市长期发展策略的一部分，其旨在转换国家的水体结构，使其超越防洪保护、排水和供水之功能，而成为充满活力、能够增强社会凝聚力的崭新城市休闲娱乐空间。截止2030年，将有100多个地点被确认阶段性实施，与已经完成的20个项目一起，将人们与水的距离拉近。公众参与被完全整合到设计过程中。

作为新加坡市中心区域最受欢迎的公园之一，碧山宏茂桥公园和加冷河修复项目成为了此总体规划之中的旗舰项目之一。与急需进行的公园升级规划同时进行的，还有公园沿线的加冷河水渠修复计划，将此单一功用性的混凝土结构河道转变为自然式河流，以创造社区居民能够充分享用的新型城市空间。

1 Drainage with green creeks
  绿植覆盖的溪流充当排水通道
2 Central watershed masterplan
  中央水环境规划

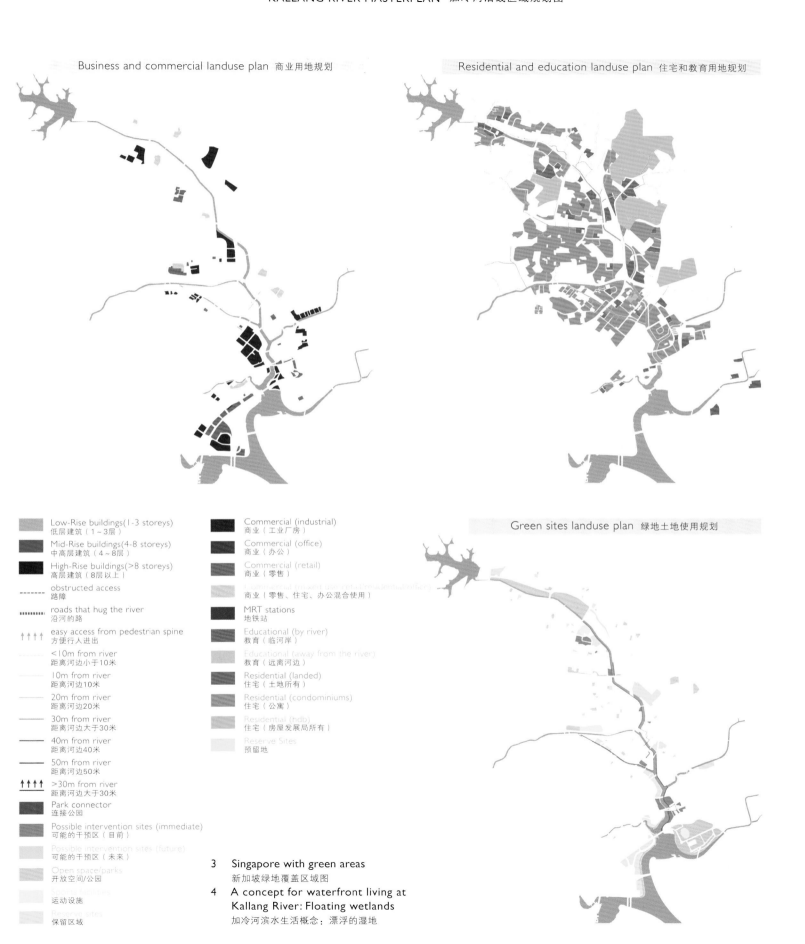

3 Singapore with green areas
4 A concept for waterfront living at Kallang River: Floating wetlands

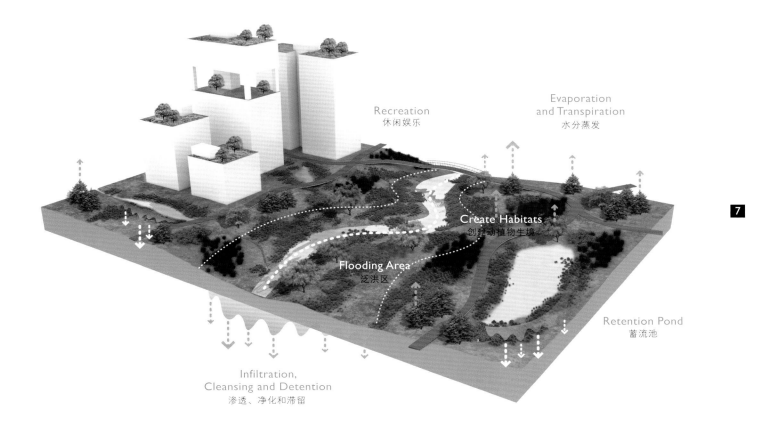

5 Reducing rain water runoff through landscape approaches
通过景观处理方式降低雨水径流量
6 Bring people closer to the waterfront
为人们提供滨水空间
7 Sustainable Integration Urban- Soft
城市软质景观可持续水体平衡
8 Cleaning of water in urban areas
城市区域的水质净化

1. ABC strategy integration of landscape, water and community
2. Water is recirculated to cleansing ponds for constant purification
3. Water quality improvement enhances the aesthectic and recreational value of the waterbodies
4. Plants irrigated using collected water
5. Stormwater collected in water bodies
6. Aquatic plants and substrate act as cleansing agents for collected water, while beautifying the environment

1. 整合景观、水文和社区环境的"活力、美观、洁净"战略规划
2. 水体不断与净化池循环实现持续的净化处理
3. 水质的提升增强了水体的美学和休闲价值
4. 利用收集雨水进行植物灌溉
5. 雨水收集
6. 水生植物和基质层充当净化主体，收集雨水并美化环境

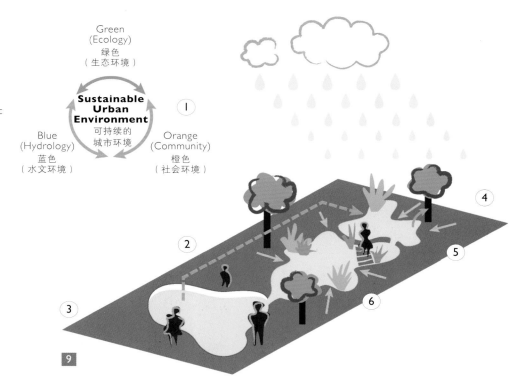

1. Wheel stop
2. Rain garden
3. Filter media
4. Transition layer
5. Drainage layer
6. Perforated pipe

1. 阻轮设备
2. 雨水花园
3. 过滤基质
4. 过渡层
5. 排水层
6. 穿孔管

1. Rock pitching
2. Filtration media
3. Street kerb
4. House drain to kerb face
5. Overflow pit
6. Footpath

1. 岩石铺装
2. 过滤基质
3. 路缘石
4. 房屋排水至路缘石表面
5. 溢流井
6. 步行路

1. Slotted kerb
2. Carriage way
3. Top of 2m wide crossing
4. Sandy loam
5. Transition layer
6. Drainage layer gravel
7. Slotted pvc pipes
8. Geotextile
9. Non-perforated connecting Pipe with connections to 600Mm wide drain
10. Weephole
11. Storm flow

1. 开槽路缘石
2. 运输通道
3. 顶部2米宽
4. 砂质壤土
5. 过渡层
6. 排水层碎石
7. 开槽PVC管
8. 土工布
9. 无孔管连接进行600毫米宽度排水
10. 排水口
11. 雨水流量

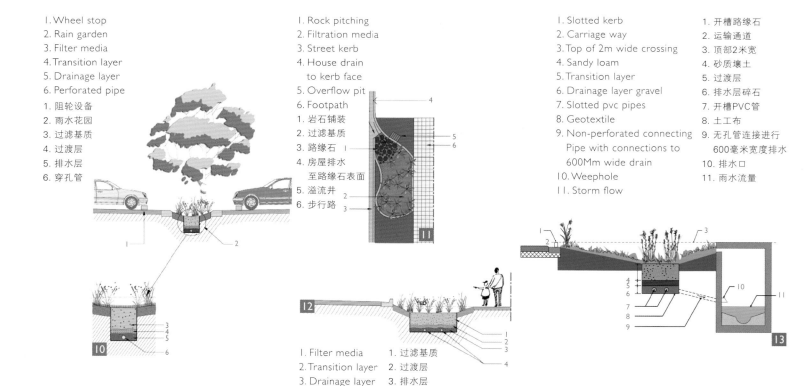

1. Filter media
2. Transition layer
3. Drainage layer
4. Perforated pipe

1. 过滤基质
2. 过渡层
3. 排水层
4. 穿孔管

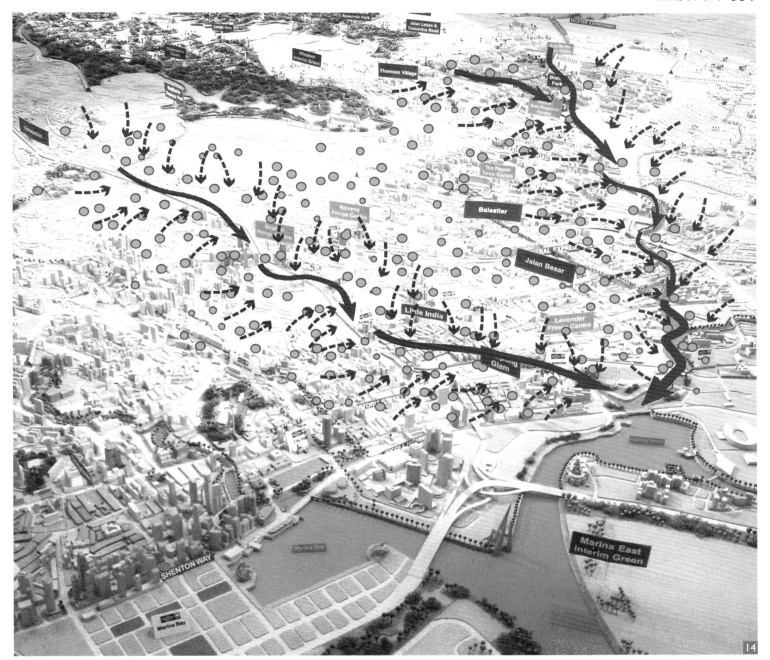

9 Integration of recreational water feature and storm water management
休闲水景特征与雨洪管理的整合
10-13 Bioretention basin and streetscape
生态滞留洼地和街景景观
14 System management of water collection
雨水收集系统管理

A02 | Masterplanning 规划设计

Tianjin Eco City
天津生态城景观规划

The determinant element of the Tianjin Eco City Site is the ancient Riverway of Ji Canal. The figure of Eco City in its general shape is defined by the scenic situation on the riverside and the bow of the old canal and by the way this setting has been interpreted by the masterplan. The Island formed by the Ji Canal and the Ancient Riverway of the Ji Canal is the Eco Core of the Eco City. This part on the isthmus between the two watercourses is reserved for research, conference and exhibition. The Eco City is built up of different Districts surrounding the ancient Riverway and the Eco Core. Each District is multifunctional and has its own Center. Green Corridors connect the Eco City with the urban areas South and East of the Site. The Eco Valley as backbone of the Eco City accompanies in a gentle S-curve the river. The building typology of Eco City though differing in its single areas has certain special aspects in common, that give this city its characteristic appearance. The buildings are offering a landscape of roof gardens, terraces, balconies and green facades and react on changing outward conditions, with facades that offer sun-protection in summer and let sunlight enter the building in cool times.

调查发现古老的蓟运河河道是天津生态城场地的决定性因素，生态城的基本形状依据河滨的风景状况以及古老运河的弓形回转而设，在总体规划之中得以阐释。蓟运河以及蓟运河古河道围合而成的岛屿形成了天津生态城的生态核心区域。两条水道之间的狭长陆地部分被保留下来，成为科研、会议举办和展览场地。天津生态城环绕古河道和生态核心区而建，设置有不同区域。每区具有多功能性，拥有其自身的中心。绿色廊道连接生态城与场地南部、东部的城市区域。生态谷作为生态城市的骨干，以一个柔和的S形曲线与河流形态相呼应。尽管生态城在单一场地范围内建筑类型不同，在总体上它们保持有一定的独特的共性，形成了城市的特色外观。建筑配有景观屋顶花园、绿化台地、景观阳台以及绿墙外观，这些建筑绿色外观共同作用以调整外部状况，在炎热的夏季遮阳，而在寒冷天气则丝毫不妨碍温暖阳光射入室内。

Location Tianjin, China
Client City of Tianjin
Partner RSAA, Rheinschiene, WLA
Area 34.2km²
Design Time 2009

项目地址 中国 天津
项目委托 天津市政府
合作机构 RSAA, Rheinschiene, WLA
项目面积 34.2平方千米
设计时间 2009年

1 Waterfront night view with the city center
水滨及城市中心夜景

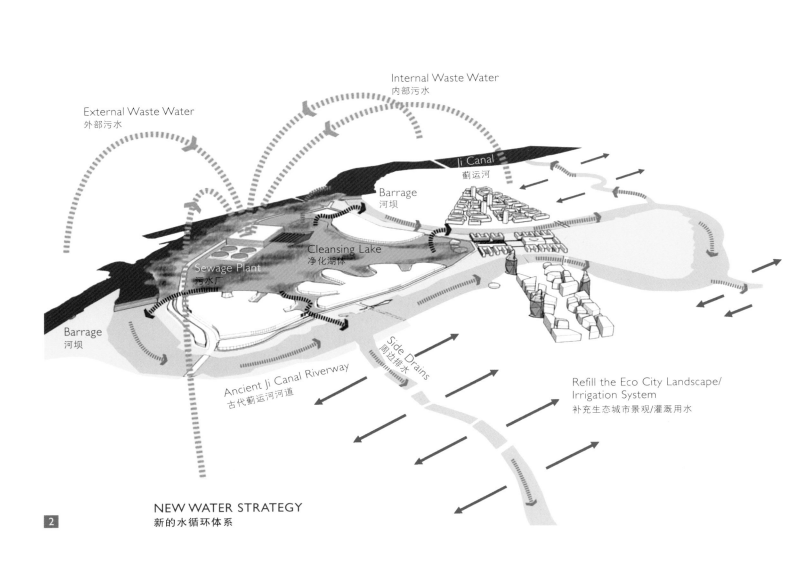

2 NEW WATER STRATEGY
新的水循环体系

2 Water management for the new City
  新城水管理概念
3 City center with riverfront
  滨水的城市中心
4 Eco towers
  生态型楼宇
5 Central area of Ecocity Tianjin
  天津生态城中心区域

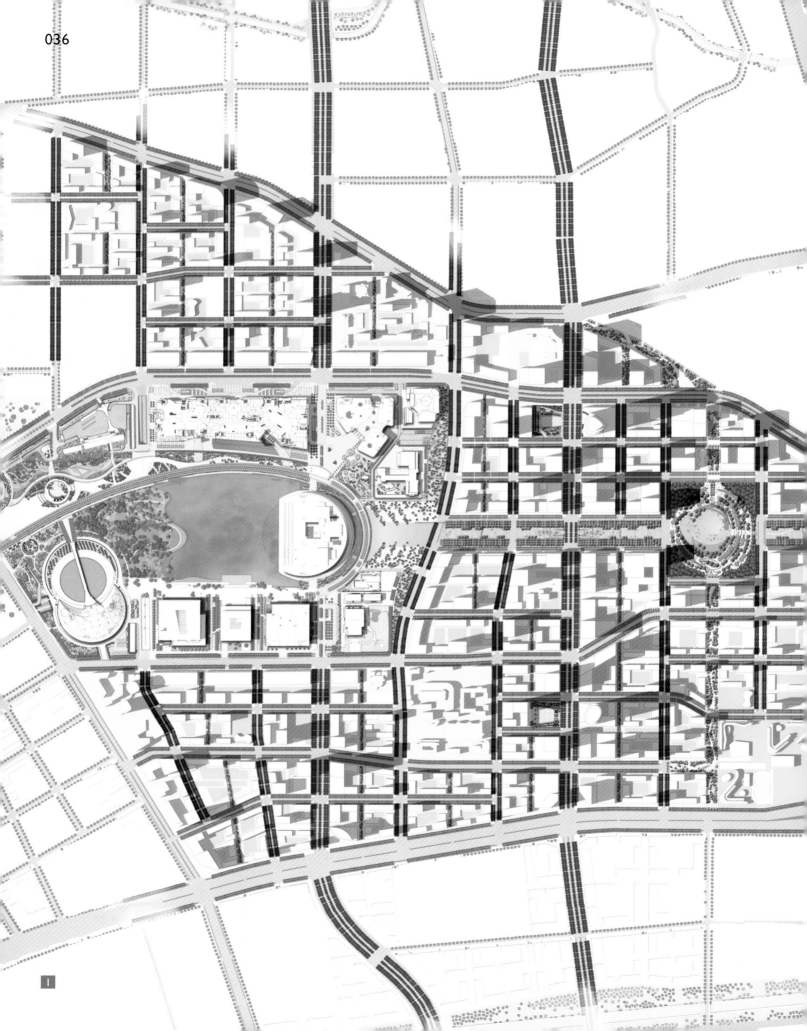

Tianjin Hexi CBD Landscape Master Planning
天津河西中央商务区整体景观规划

Location Tianjin, China
Client Tianjin Infrastructure Construction and Investment Group
Urban Design SOM
Engineer Nikken Sekkei LTD
Area 241ha
Design Time 2012
项目地址 中国 天津
项目委托 天津城市基础设施建设投资集团有限公司
城市设计 SOM
合作机构 Nikken Sekkei LTD
项目面积 241公顷
设计时间 2012年

Tianjin has historically played an important role in China as a city of global trade, rich with culture and commerce. As Tianjin evolved from a collection of historic settlements into an industrial powerhouse last century, the city is now poised for rapid growth and transformation into a forward-looking, 21st-century international center of commerce.

The development of the Tianjin Cultural Park on the south side of the city center has elevated the city's cultural offerings and created a catalyst for the development of a world class urban district. The Tianjin Cultural Center Surrounding Area Master Plan offers a sustainable framework for the redevelopment of an aging urban area into a modern business center, offering world-class office space, quality living and an abundance of public open space.

作为全球贸易城市、拥有丰富的文化和商业发展底蕴，历史上的天津在中国拥有重要的地位。20世纪，天津城市从一个历史性定居点汇集的城市演变为一个工业能源型城市，它正经历着平稳快速的增长，并蜕变成为具有前瞻性的新世纪国际商贸中心。

位于市中心南侧的天津文化中心开发提升了城市的文化给予，创造了一个世界级都市区域开发的催化剂。天津文化中心周边地区总体规划为老旧城区转化成为现代化的商务中心，创造世界一流的办公空间、优质居住和丰富公共开放空间提供了一个可持续发展的框架。

1 Masterplan with parks and green streets
公园和绿色街景规划
2 City area for transformation
待转型城市区域

Landscape Master Plan goal is to overlay complex networks of urban natural, cultural and social system over SOM's comprehensive master plan. Landscape Master Plan create visions of 'Green open space network', 'Blue Space, Water system network', 'Green Infrastructure integrated with transportation system', 'Pedestrian oriented highly usable and livable urban space network' and 'Interconnected underground spaces'. By developing these thick layers over existing master plan, TCP 2 can successfully achieve creation of world class, sustainable, and very rich and diverse networks of urban spaces.

3　View from the central park
　　从中央公园远望景致
4　Layers of infrastructure
　　基础设施层次框架

景观概念设计的目标是在区域内创建相互交叠的城市自然、文化、社会系统，并以SOM建筑设计事务所的城市设计方案作为基础框架，创建了绿色空间网络、蓝色空间和水管理网络、绿化设施与公共交通结合系统、具备高度实用性的步行导向、活力城市空间网络，以及与地下公共活动空间连接体系等城市愿景。这些网络层次在城市中相互叠加，将天津文化中心二期即天津河西中央商务区规划成为国际化、可持续、丰富多彩、充满活力的城市网络空间。

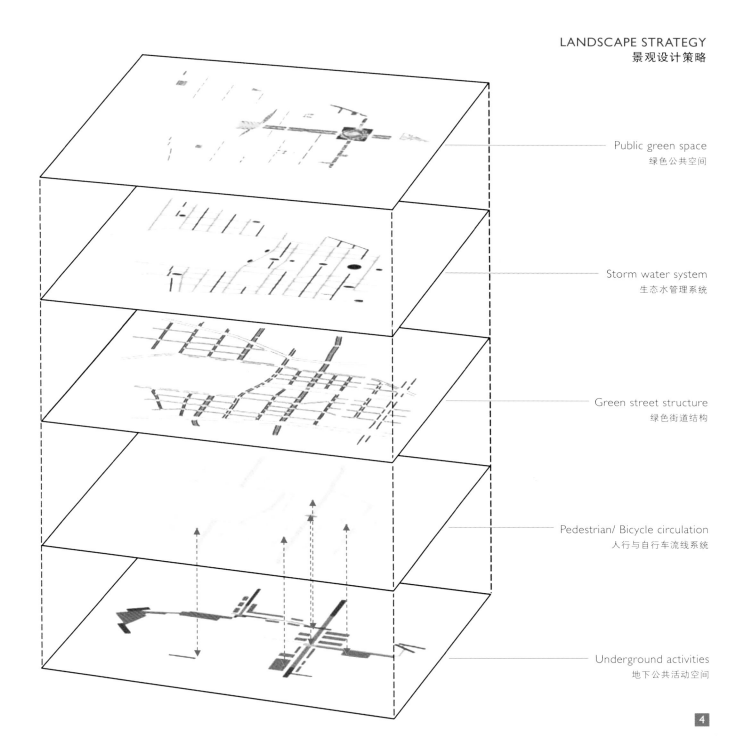

5 Outdoor comfort in streetscapes
舒适的户外街景
6 Human scale mitigation with plants
植物景色舒缓行人情绪
7 Culture park as
the heart of the City
作为城市核心的天津文化中心
8 Shading analysis
阴影分析
9 Water infrastructure, flow direction
水体基础设施，水体流动方向

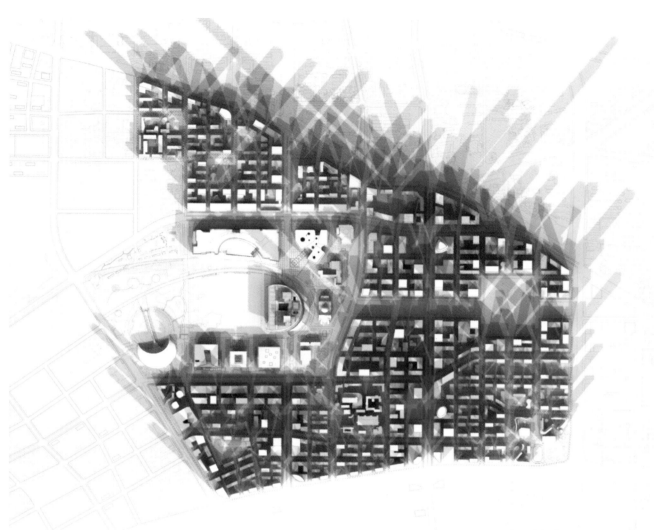

LANDSCAPE STRUCTURE 景观结构

The planning area
规划范围
The production protective green belt
生产防护绿地
The public green space
公共绿地
The green space at the historical site
文物保护绿地
The artificial wetland
人工湿地
The natural ecological wetland
自然生态湿地

2

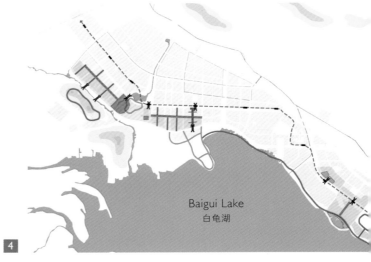

The planning area
规划范围
The landscape belt
核心景观带
The key landscape belt
景观锲入绿带
The park landscape joint
公园景观节点
The commercial landscape joint
商业景观节点
The wetland landscape joint
湿地景观节点
The sports landscape joint
运动景观节点
The lakeside landscape joint
亲湖景观节点
The mountain landscape joint
山地景观节点

3

4

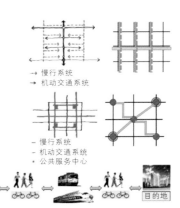

The planning area
规划区
The featured commercial area
特色商业街区
The business streets
商业步行街
The commercial pedestrian street
休闲商业街
The walking trails
步行道
The hiking trail
自行车道
The overpass
过街天桥
The metro line
轨道交通线
The metro station
轨道站点

1 Sport and recreation area
   体育和娱乐区域
2 Green system
   绿地系统
3 Landscape system
   景观系统
4 Transportation pattern
   交通方式

Development Masterplan, Pingdingshan, Henan, China
河南平顶山生态景观与水文发展规划

Location City of Pingdingshan, Henan Province, China
Client City of Pingdingshan
Urban Design Pesch Partner Architekten Stadtplaner
Area 2,872ha
Design Time 2011
项目地址 中国 河南省 平顶山市
项目委托 平顶山市
城市设计 德国佩西规划建筑设计事务所
项目面积 2872公顷
设计时间 2011年

The key design elements - Nature, City and Water - just like Chinese classical artistic ribbons interweave and mingle mutually. Combining ecological and geometrical measures, this masterplan aims to create the unique landscapes and harmonious eco-city by the chance of urban development.

The main design principles are remembering water's source when drinking and studying from Nature. Rain water soaks firstly into the ground. During this process, rain water is cleansed through infiltration, being stored and used as resource and treasure for all creatures. Water run-off feeds rivers and waterways by following the natural terrain. The lower land works as wetland to give the water a detention space where it can get cleansed at the same time. Due to the slow motion of rain water drainage, it creates a lively space for nature and human beings.

Learning and following the rules of nature, this project is creating a new city in harmony between people and nature. Landscape element of master plan consists of hills, lowland and water bodies with its natural functions. These elements are integrated together into green structure of future eco city. Water bodies offer buffer space during heavy rain events. The excellent and reasonable usage of water resource is core part of this ecological landscape design.

自然、城市、水——三个最重要的元素犹如中国古典艺术的飘带，相互映衬，相互依托。该设计利用生态、有机的组合方式，利用城市发展的机会，搭建平顶山特有的景观特质，形成和谐的生态城市。

主设计原则：饮水思源，师法自然。水是万物之源。降水首先浸润大地。在此过程中，雨水得到渗透和过滤，蓄存之后成为惠及万物的资源。地面的径流则顺着天然的地势汇入河流和水道。低洼区域则作为滞留雨水的湿地。得益于这样一个缓慢进行的雨水沉降过程，自然和人类得到了一个和谐生机的空间。

继承现有景观，遵循自然法则，将其运用到我们的项目中，创造人与自然的和谐共处。景观规划要素包括山体、低地以及拥有自然功能的水体，它们被精细地布置和整理，构成未来生态城市的绿色肌理。暴雨来临之时，水体提供蓄洪和削减洪峰的空间。对于水资源高效、合理的利用是此生态景观设计之核心部分。

5 **Stormwater lake as protection and feature**
   雨水湖在作为景观的同时可起到保护作用
6 **Lake elevaltion from main lake**
   相对于中心湖的湖体高度
7 **Urban structure shaping**
   城市结构图
8 **Stormwater collection to the lake**
   雨水汇集入湖

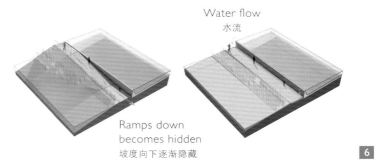

Water flow 水流
Ramps down becomes hidden 坡度向下逐渐隐藏
6

Roads adjustment in Eastern area: from curl to straight line and open to the lake; lake running along edge of city to form exciting space (Below).
东片区道路调整：化曲为直，向湖打开；城湖穿行，趣味盎然

Design concept of roads in Eastern area
东片区道路调整概念

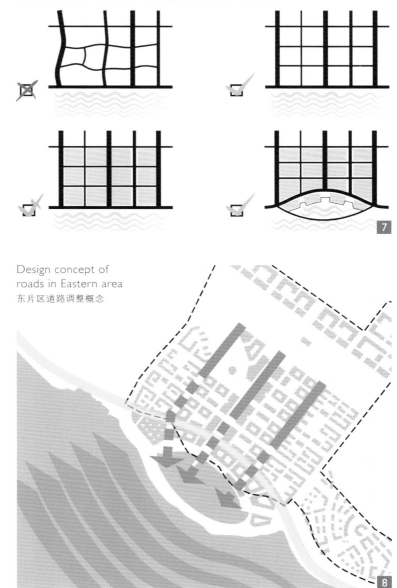

046

Duisburg 杜伊斯堡
Oberhausen 奥伯豪森
Gladbeck 格拉德贝克
Bottrop 波特洛普
EMSCHER 埃姆舍河
Lanferbach 兰弗巴赫
Gelsenkirchen 格尔森基尔欣
Herten 黑尔腾
Recklinghausen 雷克林豪森
Waltrop 瓦尔特罗普
Castrop-Rauxel 卡斯特罗普-劳克塞尔
Info-Zentrum 信息中心
Herne 赫尔内
Westfalen-Stadion 威斯特法伦球场
Bochum 波鸿
Essen 埃森
Mülheim A.D. Ruhr 米尔海姆
RHEIN-HERNE-KANAL 莱茵-赫尔内运河

1

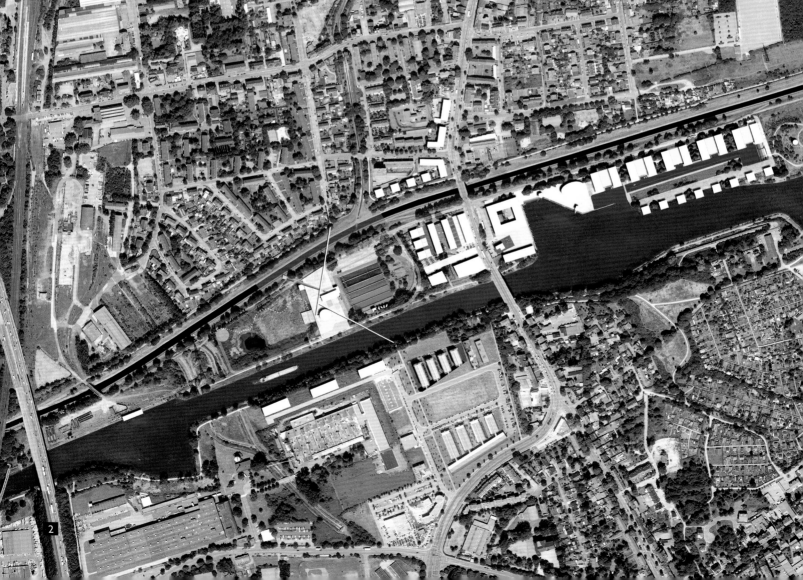

2

Restoration of Emscher River, Ruhr Valley, Germany
德国埃姆舍河流域修复规划

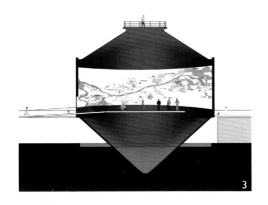

Location Ruhr Valley, Germany
Client Emscher Genossenschaft
Partner Schaller/Theodor Architects
Area 15ha
项目地址 德国 鲁尔区
项目委托 埃姆舍项目合作组织
合作机构 Schaller/Theodor Architects
项目面积 15公顷

1. Watershed emscher river
   埃姆舍河流域图
2. Old plant for a new exhibiton center
   拟建新展览中心的旧场址
3. Showroom with projection of the emscher
   埃姆舍投影陈列室

The Emscher master plan is ambitious; the plan of the 50 km long Emscher river corridor should be reworked and by 2020, the impervious surfaces in the rain water collection area must be reduced by 30%.

In order to give the long-term strategy a coherent workflow, we were charged with developing concepts for 14 projects. Each project was very different, for example, altering paving surfaces in a 1920's residential area to integrate a rain water system or changing a former settling basin into an outdoor swimming pool.

However, the most important project was to produce new concepts for environmental and water management, for the present generation to take forward. The Emscher Info Park was developed to raise public awareness and increase knowledge and involvement.

Set on an island between the Emscher and a navigable canal, an information and activity centre will be integrated to an old digester in a former sewage plant, whilst old sewage basins will illustrate water phenomena and rainwater cleansing processes.

埃姆舍总体规划目标远大，50km长的埃姆舍河走廊将被翻新，截止2020年，雨水收集区域的不透水地面需减少30%。

为了将此长期战略项目形成一个连贯的工作流程，按要求我们共规划设计14个项目的发展概念。每一个项目都各自不同，例如改变20世纪20年代住宅小区内的路面铺装、将之整合成为雨水收集系统，或将之前的沉沙池改造成为一个室外游泳池。

然而，最重要的环节是开发环境和水资源管理的新概念，并继代推进。埃姆舍信息园区的开发，即以提升公众意识、普及知识和增强参与为目标。

埃姆舍和通航运河之间的一个小岛上，一个信息和活动中心将被整合为一个旧污水处理厂的沼气池，而旧的污水池将展示水体现象和雨水净化处理过程。

A06 | **Masterplanning** 规划设计

Offenbach Harbour, Offenbach, Germany
德国奥芬巴赫港口区域规划

An industrial peninsular on the River Main is being conver ted into a new sustainable city district. The "harbour" forms the new urban focus with varied access to the water and multifunctional urban space. A promenade network connects the water to the more densely build-up area on one hand and to the new local park and regional river path beyond. In the challenging environment of contaminated soil, a holistic stormwater management concept has been developed to create soft city spaces and streetscapes while retaining and cleansing stormwater before releasing it to the river and harbour. Innovative natural water treatment systems such as cleansing biotopes are integrated into the park spaces, and new natural habitats are created for riparian flora and fauna whilst proving refreshing green oasis for the city. This project is a pragmatic vision for how climate adaptation can be achieved in urban context creating attractive neighbourhoods through integrated engineering. The project has prequalified for the prestigious DGNB Gold for a sustainable city district, comparable to a LEEDS Platinum rating, a European first.

美茵河上的一个工业半岛被转换成为一个新型可持续城市区。"海港"形成新的城市焦点，通过多样的途径通往水边和多功能的城市空间。散步道网络系统一方面连接水体与密集的建成区域，同时连接新建地区公园和公园范围之外的地区河流道路。在土壤污染严重、充满挑战的环境中，一个全面的雨水管理概念被付诸实践，以创建柔质城市空间和街景，同时在雨水汇入河流和港湾之前，对它们进行存续和净化。净化群落等创新型天然水处理系统被整合入公园空间之中，新的自然栖息地被创造，为河岸边动、植物提供栖息地，同时也为城市创造了一处清新的绿洲。该项目为在城市背景之下通过综合的工程手法、创建富有吸引力的街区、以适应气候变化提供了一个务实的解决方式。该项目被授予相当于LEED铂金奖的欧洲第一的DGNB可持续城市区域金奖。

Location Offenbach Harbour
Client Mainviertel Offenbach GmbH & Co. KG
Infrastructure and Transportation Plan Schüßler-Plan
Urban Plan Berlin Ortner + Ortner
Area 25.6ha + 6ha catchment area
Design Time 2008-2019
Construction 2009-2020
Awards DGNB Sustainable City District Gold Award, 2011

项目地址 奥芬巴赫港
项目委托 Mainviertel Offenbach GmbH & Co. KG
基础设施与交通规划 Schüßler-Plan
城市规划 柏林Ortner + Ortner公司
项目面积 25.6公顷+6公顷水域面积
设计时间 2008年–2019年
施工时间 2009年–2020年
所获奖项 2011年DGNB可持续城市区域设计金奖

1　Main plaza with a staircase to the water
　　中心广场，拥有台阶通向水面

A06 | Masterplanning | 050

2　Old crane in the former industrial harbour
　　开发前工业港口上的老旧起重机
3　Staircase looking out to the water
　　遥望水面的阶梯景观
4　Plan for the mix use development
　　多功能开发计划
5　Staircase with seating
　　附有座椅的阶梯景观
6　Boardwalk at the old river bank
　　河边的步行道
7　Vision for the new city
　　新城远景

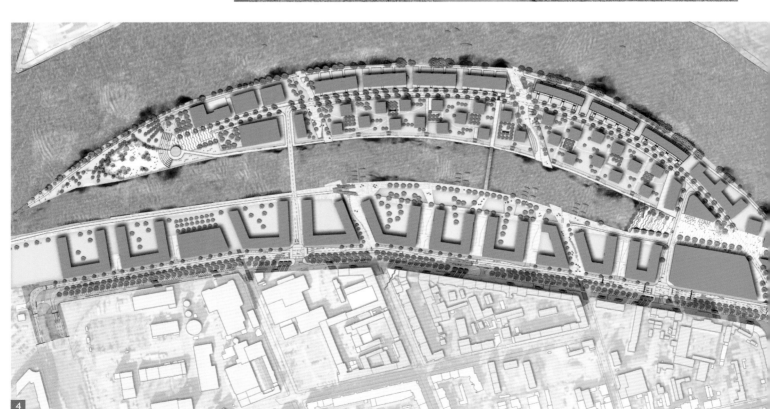

Eco Quarters – How to Build Truly Sustainable?
by Gerhard Hauber

To design a city or a city quarter is a highly complex task. Usually based on a feasibility assessment and projection about what is economically and socially needed a design team starts to develop a masterplan. The architect starts to layout the urban structure, the engineers design energy, traffic and other infrastructure concepts and the landscape architects design the green spaces. The working method is known and proven in many projects and so, many of these new developments are done in a traditional way. The focus is usually that the difference of investment and return is as high as possible. The result financially may be a success but the result for our environment and the dramatic consumption and loss of finite resources is dramatic. Worldwide 30 to 40% of the energy consumption is related to buildings and approx. 40% of all Green House Gas emissions is produced by the building sector[1].

We also know that the future of cities in many regions is "hot", which means the mean temperature in the summer will rise dramatically in the next 30 years while rain events will be less frequent but much more heavy. This has tremendous impacts as for example our stormwater infrastructure is not able to handle these changes and the rising temperature will create serious health problems, increase energy consumption and air pollution. It is more than obvious that a dramatic change in the way how we develop urban districts is necessary. The good news about these problems is that we are aware about them and that we can change it with relatively moderate efforts. There are enough pilot projects done that can show us what could be a sustainable solution.

Important is that we get an answer to the questions: What is a sustainable City? How are we measuring sustainability? What is needed to really make a radical change? A new development can be created with all good intentions to be as resource protective and sustainable as possible but how do we know that this is the truth? What can guide us through a complex design and development process?

There is not a perfect answer but the best way to find this out currently is certainly to use a rating system. When used consequently and concurrently while the project is developed it can guide the decisions towards true sustainability. It can help to find out the most efficient aspects related to sustainability, so that the efforts and focus for a specific project in a specific natural, climatic and urban environment is used in the best way possible. A good example is the new rating system from the DGNB (German Sustainable Building Council) for City districts. It is a second generation rating system with a very holistic approach than for example the LEED-System. Only the DGNB-System considers the costs of the complete life-cycle, only the DGNB-System factors the economic costs and also the technical quality is not considered in the other systems.

This holistic assessment makes it to the best system currently available. The main goals are the protection of the environment and natural resources, the minimization of life cycle costs, the increase of health, comfort and well-being of the citizens, sustainable mobility and the promotion of the adaption of innovative concepts and technologies. Based on these goals the system analyzes five major aspects of the projects: the environmental quality, economic quality, socio-cultural and functional quality, the technical quality, and process quality.

(1) http://www.eco-business.com/blog/energy-efficient-buildings-green-or-grey-green/

Each of these main topics is separated into 3 to 12 sustainability criteria, which totals to about 50 criteria. Here the individual aspects are analyzed and evaluated against current benchmarks. So for example the life-cycle-costs are considering the installation costs, the maintenance efforts, as well as the repair costs throughout the expected life-cycle. In the so called "eco-balance" criteria the environmental impact of each material is analyzed. In the biodiversity criteria is the potential of the site for the survival of local species of flora and fauna analyzed and how the development is connected with biotopes in the surrounding.

The urban open space plays an important role in this system. More than 30% of the criteria are related to the open space and the integration of state-of-the-art blue-green infrastructures. The open space will play a key role in the mitigation of the impacts from the climate change. Open space is now available, it can be designed multi-functional (when dry used for people – when it's raining used for filtration, retention and evaporation of storm water) and if the creation of a connected blue-green citywide network is the focus for the next decade it will keep our cities alive and livable.

The assessment of the first DGNB-audit processes has proven that this rating system can guide a planning and development process towards sustainability. While doing the assessment parallel to the design process all decisions are done on a scientific basis and with clear knowledge about the effects on the sustainability of the project.

Interesting is that one of our more than 15 year old pilot projects, the Potsdamer Platz in Berlin, Germany has been audited last year. It was rated Silver which is astonishing for such an old project. The main factor for getting rated so positive was the urban open space with the holistic water concept. A pilot project in China, with the same philosophy was realized in Tianjin with the open space and sustainable water system of the Cultural Park in the city center.

Again the question comes back: What would have been possible with the knowledge and guidance of the DGNB rating system?

We are sure, when used consequently the benefits are immense, not only that the investor gets a high quality project that keeps his property value high and which is validated by the rating system, it is also created with the goal to realize a modern, long lasting and socially stable development which respects and supports the citizens. We see an increasing need to verify sustainable urbanization in Asia and specifically in China, because that is the only way to change from a fast growing unsustainable urbanization to a quality-based long-lasting growth.

生态区域——如何建立真正的可持续发展？

格哈德·豪博

设计一个城市或城市街区是一项非常复杂的任务。通常，基于对于可行性的评估和经济、社会需求的预测，设计团队开始制定一个总体规划。建筑师开始布局城市结构，工程师开始规划能源、交通以及其他基础设施概念，景观设计师们则开始设计绿色空间。如此的工作方式被大家熟知，在众多项目之中获得验证，因此，许多项目的开发都以这样的传统方式开展和实践。关注点通常被放在了投资和回报之间尽可能悬殊的收益之上。最终，财务方面的结论也许说明了项目的成功，但对于环境和地球有限的资源的巨大消耗和损失却是剧烈的。全球范围内30%~40%的能量都消耗在建筑物及相关配套设施上。全球40%的温室气体排放量来自建筑物[1]。

我们也知道，未来许多区域的城市将会成为"热点"地区，这意味着在接下来的30年里，夏季的平均气温将急剧增长，而降雨将变得不那么频繁、但却更加剧烈。如此便会对于比如雨水基础设施带来巨大的影响，它们将无法应对这些变化。温度上升会引发严重的健康问题，增加能源消耗并加剧空气污染。因此很明显地，我们开发城市方式的根本性转变已成为了必需。而应对这些问题的好消息便是，我们已意识到、并且能够采取相对温和的方式来改善。足够多的试点项目已经被实践，向我们展示了一种可持续的解决方案。

重要的是，我们需要获得答案，以解答如下问题的：什么是可持续发展的城市？我们该如何实现其可持续性？我们需要做些什么以真正实现根本性的转变？尽管我们可以创建一个拥有一切美好愿景的新发展道路，并尽可能地保护和维持资源的可持续性，不过我们又怎么知道这就是真理？而又是什么知识能够对我们复杂的设计和开发过程进行指导？

尽管没有一个完美的答案，但目前最好的方式无疑是使用评级系统。在项目开发的同时应用该系统进行评判，它便可以对于方案和决定进行指导，以实现真正的可持续发展。这种系统可以帮助找出有关可持续发展的最有效环节，这样，在一个拥有特定自然、气候和城市环境中的具体项目中，各种投入和付出便以尽可能最好的方式得到应用。在这里，德国可持续住宅委员会（DGNB）制定的城市地区全新评级系统便是一个很好的示范。它是一个比较如LEED评级系统更为全面的第二代评级系统。DGNB系统考虑完整的生命周期成本，并同时考虑经济消耗与技术质量，这在其他评级系统之中还不曾涉及。

这种全面的评估使它成为了目前最好的系统，其主要目标是保护环境和天然资源，最大限度地降低生活成本，提高公民的健康水平、生活舒适度和福祉，增强可持续的交通流动，并提升创新观念和技术的适应性。根据这些目标，该系统对于项目的五大方面进行了分析：环境质量，经济质量，社会文化和功能质量，技术质量以及过程质量。

(1) http://www.eco-business.com/blog/energy-efficient-buildings-green-or-grey-green/

这些主要方面之中的每一项都包含3~12个可持续发展的标准，总计约50个标准。如文中图表所示，针对每一个方面进行了分析，并对当前的基准进行了评估。例如，对于生命周期成本的考虑包括安装成本、维护投入，以及贯穿整个预计生命周期之内的维修成本。在"生态平衡"的标准之中，分析了每一种材料对环境可能产生的影响。而"生物多样性"标准则是指项目场地对于本地动植物物种生存的潜在影响以及该开发项目如何与周围生境连接。

城市开放空间在这个系统中发挥着重要的作用。30%以上的标准都针对于开放空间以及与艺术表现形式有机整合的蓝绿基础设施。开放空间将在缓解气候变化的影响方面发挥关键的作用。开放空间是现实存在的，它可以被设计成为具备多重功能（当晴朗干燥时作为绿色空间为人们使用，而在下雨之时则作为雨水过滤、滞留和蒸发的场地）。如果能够将创建一个遍布城市范围、贯通连接的蓝绿网络作为未来十年之内城市建设的关注焦点，它将使得我们的城市充满活力、变得宜居。

对于DGNB审核过程的第一轮评估已经证实，这种评级系统可以引导规划和发展过程以实现可持续化。当这样的评估与设计过程并行进行时，所有决策的制定便能够基于科学的依据，并可以明确知晓具体的设计对于项目可持续性的影响。

值得一提的是，由德国戴水道设计公司进行雨水收集、管理设计的、建成已经超过15年的试点项目——德国柏林波茨坦广场，在去年通过DGNB的审核。令人兴奋的是，这样一个已建成十余年的"老"项目获得了DGNB银奖。获得如此高评价的最主要因素就在于该项目之中所体现的城市开放空间的整体水概念。相同理念的设计在位于中国天津城市中心区域的天津文化中心项目的开放空间和可持续水环境设计之中同样得到实践。

问题又回来了：DGNB评级系统的知识和指导能够带来什么？我们相信，如果能够合理地加以使用，获益是难以估量的，不止投资者能够获得高品质的项目，通过评级系统认证将他的房产价值提高，同时它也建立了一个现代、长期、尊重并有益于市民的社会稳定发展。我们完全可以清楚地看到亚洲地区尤其是中国对于可持续性城市化的需求，因为这是将之从一个快速增长的不可持续性城市化转变为以质量为基础的长久持续增长模式的唯一途径。

Ecological Infrastructure & Stormwater Management in Tianjin Cultural Park
天津文化中心生态水环境基础设施

Location Tianjin, China
Client Tianjin Environmental Construction Investment Co., Ltd
Landscape planning and water design Atelier Dreiseitl
Architects GMP, KSP, Riken Yamamoto, HHD, Callison, ECADI, TVSDESIGN
Partner WLA, TADI, YMGC
Engineer Polyplan
Area 90ha
Design Time 2009-2011
Completion Time 2012
Award 2012 Certificate Honor, Outstanding Design

项目地址 中国 天津
项目委托 天津市环境建设投资有限公司
整体景观规划及水环境设计 德国戴水道设计公司
建筑设计 GMP, KSP, Riken Yamamoto, HHD, Callison, ECADI, TVSDESIGN
合作机构 WLA, TADI, YMGC
工程顾问 Polyplan
项目面积 90公顷
设计时间 2009年-2011年
建成时间 2012年
所获奖项 2012天津市杰出设计奖

1 Central Lake as a focus for sustainablility
作为可持续设计焦点的中心湖

Tianjin is one of China's top cities, not just in size and population but also in terms of business investment. Located just half an hour south-east of Beijing by high-speed train, Tianjin is also close to the sea. The high-water table needs to be maintained to prevent seawater encroaching inland and the dry, harsh climate does not preclude flooding. In the design of the new cultural district between the new opera house and existing city hall, a main goal was to increase outdoor comfort and create dynamic, social pedestrian routes. The lake waterfront is aesthetic with dramatic views to the opera house and exciting Museum, gallery and library frontage. Avenues of trees and planting shield the waterfront from the cold Mongolian winds while at the same time storing water for irrigation.

The lake is a stormwater feature, a balancing pond which can effortlessly handle a 1 in 10 year storm event and buffer a 1 in 100 storm event. Generous tree plantings link subsurface, decentralised retention trenches which feed the lake via a cleansing biotope. The urban lake has its own natural biology and reduces temperature extremes. Its scenic beauty sets the scene for Tianjin's most outstanding new cultural architecture.

作为中国重要的大城市之一，天津在城市面积、人口规模以及商业投资方面都占据着重要的位置。位于北京东南部、拥有仅半小时高铁车程，天津是一座海滨城市。因此，需要维持较高的水位线，以防止海水向内陆倒灌；恶劣的干燥气候，并不能阻止洪涝灾害的侵袭。在新建大剧院和现存市政厅之间，一处新的文化区域被设计建造，其主要目的为增强场地室外的舒适性，同时创造充满活力的社交步行线路。隔湖望向大剧院、博物馆、艺术馆和图书馆，变幻的景致极具美感。行道树列和种植植被对于寒冷的蒙古冷风起到了一定的屏蔽作用，同时有利于灌溉雨水的蓄积。

湖体具备雨水管理的功能，蓄积池塘可以毫不费力地应对10年一遇的暴雨事件，并对100年一遇洪涝灾害起到缓冲作用。大量的树木种植连接地表下分散的雨水滞留通道，收集雨水在经过生态净化群落的净化作用之后补给湖体用水。天津文化中心内的湖体拥有其自身的生态特征，能够发挥降低炎热天气极端高温的作用。秀丽的景致使得文化中心成为天津城区最为突出的新文化区域。

B01 | Ecocity District | 058

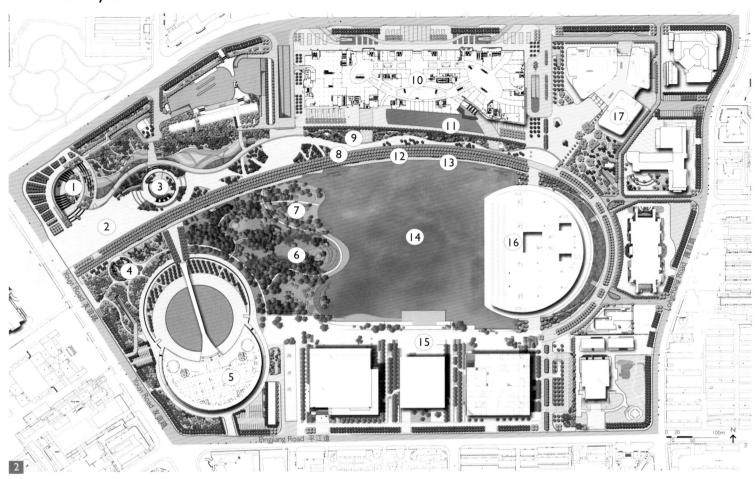

2　Masterplan
规划方案

3　Water management concept
水体管理概念

4　Green, circulation, public transport, energy
绿地、流通、公共交通、能源规划方案

5-9　Parks, plazas, promenades, small and big scale
公园、广场、步行道以及多尺度的开放空间

Legend:
1. Corner Sunken Plaza
2. Galaxy Plaza
3. Central Sunken Plaza
4. Garden Sunken Plaza
5. Museum
6. Eco Island
7. Cleansing Biotope
8. Plantane Promenade
9. Promenade Garden
10. Shopping Mall
11. Mirrorlike Water
12. Malus Speetabilis Promenade
13. Waterfront
14. Central Lake
15. Cultural Terraces
16. Opera
17. Youth Centre

图注：
1. 角部下沉广场
2. 银河广场
3. 中央下沉广场
4. 花园下沉广场
5. 博物馆
6. 生态岛
7. 生态净化群落
8. 法桐步道
9. 步道花园
10. 商业建筑
11. 镜面水池
12. 海棠步道
13. 亲水驳岸
14. 中心湖
15. 文景台
16. 大剧院
17. 青少年中心

Circulation and Refill
循环与补水方案
Water Supply Proposal
调水（区域水循环）方案

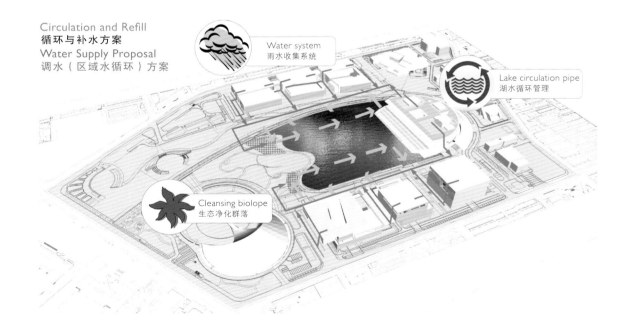

Energy System
能源系统

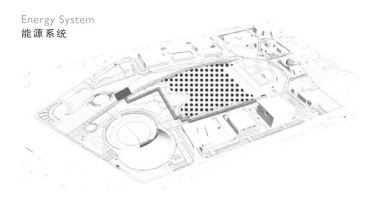

Public Transport Subway
公共交通网
Regional Transportation Cycle
区域交通循环

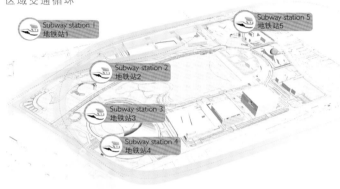

Pedestrian Circulation
道路循环系统

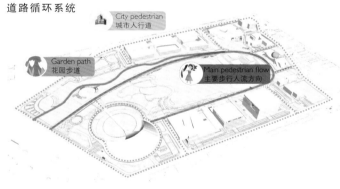

Green Volume
城市绿化系统
Main Traffic Flow
主要交通流线

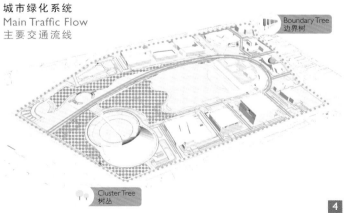

生态区域设计 | 059

10 Overview with all buildings
规划鸟瞰图
11 Cultural buildings by night
文化建筑夜景图
12 New district in the existing city
城市之中的新建区域
13 Opera building as the center
作为场地中心的大剧院

生态区域设计 | 061

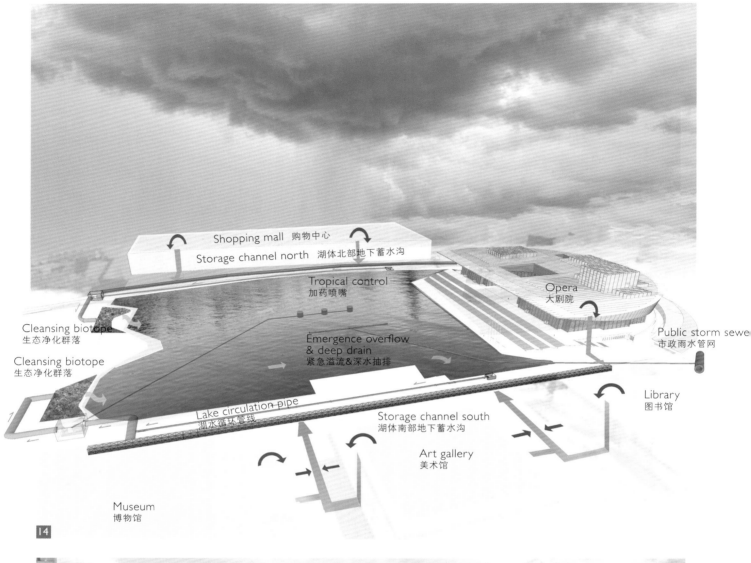

14

15

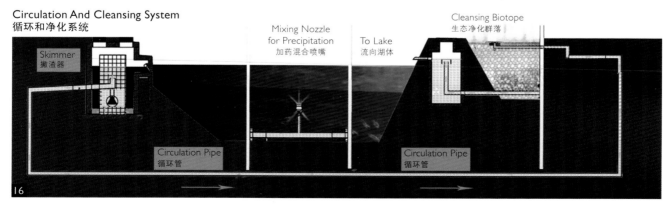

16

14 Stormwater collection, storage and treatment
雨水收集、存储和处理

15 Stormwater underground storage
雨水地下收集

16 Water treatment and Lake refill
雨水处理和湖水补充

17 Lake topography for an optimal limnology
能够发挥最佳湖体净化功能的形态设计

18 Heavy rain event fills the buffer of the lake
遭遇暴雨时雨水溢满湖体

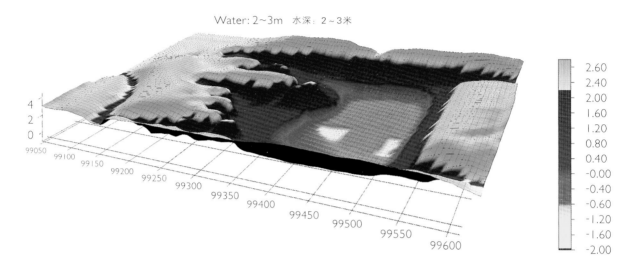

Water System Potsdamer Plaza, Berlin, Germany
德国柏林波茨坦广场水系设计

Location Berlin, Germany
Client City of Berlin, Daimler Chrysler Immobilien
Partner Renzo Piano Architects, Kohlbecker Architects
Area 1.3ha
Completion Time 1998
Awards 2011 DGNB Silver Sustainable urban district

项目地址 德国 柏林
项目委托 柏林市戴姆勒克莱斯勒公司
合作机构 Renzo Piano Architects，Kohlbecker Architects
项目面积 1.3公顷
建成时间 1998年
所获奖项 2011年DGNB可持续城市区域设计银奖

The iconic Potsdamer Plaza bridges the scar left by the wall between East and West Berlin. A veil of shallow flow-steps create a rhythmic surface of shimmering waves, providing multiple opportunities for people to cross and interact with the water. This urban waterscape has contributed to make Potsdamer Plaza one of the most visited places in Berlin.

The idea behind this important urban waterscape is that the rainwater should be used where it falls. At Potsdamer Plaza, a combination of green and nongreen roofs harvest the annual rainfall. Rainwater then flows through the site's buildings and is used for toilet flushing, irrigation, and fire systems. Excess water flows into the pools and canals of the outdoor waterscape creating an oasis for urban life. Vegetated biotopes are integrated into the overland landscape and serve to filter and circulate the water that runs along streets and walkways, all without the use of chemicals. The lake's water quality is excellent forming a natural habitat and fresh water usage in the buildings has been reduced. Since 1998, Potsdamer Plaza stands as a successful example of a revitalized open space where city life, prestigious architecture, and the beauty of water are in harmony.

标志性的波茨坦广场承载着东、西柏林分裂而遗留的历史创伤。如薄沙般浅浅的流动台阶在微风拂动下，形成波光粼粼的韵律表面，为人们提供更多的亲水、戏水乐趣。此城市水景设计使得波茨坦广场成为柏林著名的游览场所之一。

这一城市水景设计之中蕴含的理念即为，雨水在降落之地即被就地使用。在波茨坦广场，绿化屋顶和非绿化屋顶的结合设计可以获取全年降雨量。雨水从建筑屋顶流下，作为冲厕、灌溉和消防用水。过量的雨水则可以流入户外水景的水池和水渠之中，为城市生活增色添彩。植被净化群落融入到整个景观设计之中用以过滤和循环流经街道和步道的水质、水体，而无任何化学净水制剂的使用。湖水水质很好，为动植物创造了一个自然的栖息场所。同时，由于净化雨水的再利用，也使得建筑内部净水使用量得以减少。自1998年建成之后，波茨坦广场已经成为一个开放空间重获生机的成功案例。在这里，城市生活、杰出的建筑创作和魅力的水景实现了和谐统一。

1 Marlene Dietrich Plaza with the casino
玛琳黛德丽广场和赌场

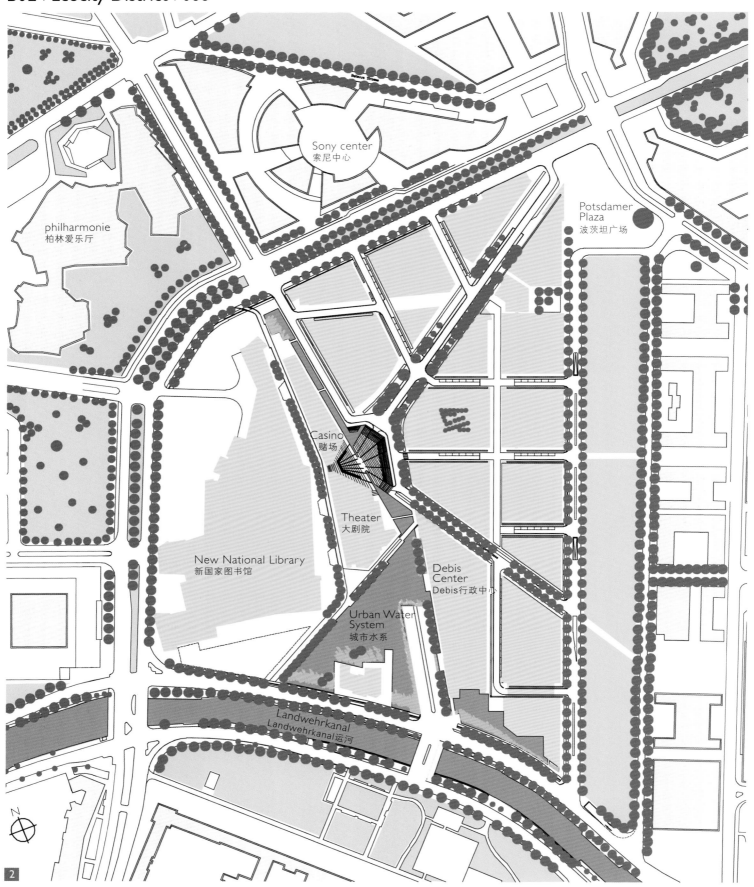

2　Plan of the Potsdamer
　　Plaza development
　　波茨坦广场发展规划
3　Seating steps
　　attracting many people
　　吸引行人的可乘坐阶梯平台
4　Recreational value
　　in midst of the dense city
　　密集城市之中的休闲场地价值
5　Water wraps along
　　the pedestrian areas
　　沿步行区域的汇水区

6 Urban lake with view to the center
   城市中心区域的河流景观
7-9 Close connection between building and water
   建筑与水体的紧密连接
10 Overall sustainable water concept
   整体可持续性水概念
11 Water treatment with natural cleansing biotopes
   以自然净化群落进行雨水净化处理

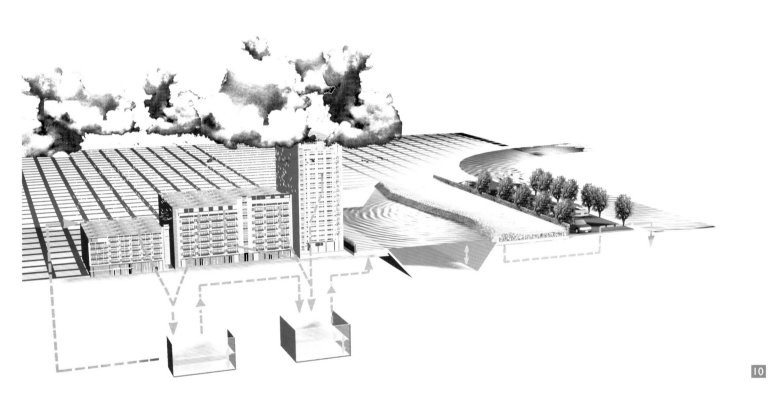

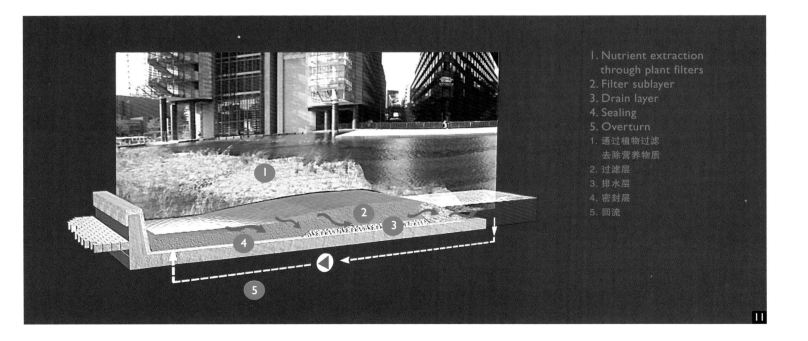

1. Nutrient extraction through plant filters
2. Filter sublayer
3. Drain layer
4. Sealing
5. Overturn

1. 通过植物过滤去除营养物质
2. 过滤层
3. 排水层
4. 密封层
5. 回流

Transformation of a City District, Taipei, Taiwan

台北城市街区的转变

Location Taipei
Client Far Eastern Group
Partner EDS Architects, Taipei;
Theodor/Schaller Architects
Area 24ha
Design Time 2007-2011
Construction Time 2008-2012

项目地址 台北
项目委托 远东集团
合作机构 EDS Architects, Taipei、
Theodor/Schaller Architects
项目面积 24公顷
设计时间 2007年-2011年
施工时间 2008年-2012年

Atelier Dreiseitl was commissioned to create a new city vision for Taipei. A humid subtropical climate subject to typhoons makes flooding a common occurrence in the city while the local Keelung River has been seriously degraded. An annual rainfall of 2000mm has in recent years been augmented by typhoons which add the same volume of water again in a few days. Heavy clay soils compound the challenge of working with rain in-situ in this dense city. Rescuing the almost extinct local paddle fish became the byline for a project which creates fresh and lively public realm which intrinsically manages rain as a city resource on a local level.

A water sensitive urban design highlight is the tree avenues which connect subsurface retention channels supplying the nearby Keelung River with clean water. A central park brings a breath of fresh air to the city and an urban lake balances typhoon rains. 50% of rainfall is managed on site and a flood strategy accommodates a 1 in 100 year flood event. Pedestrian streetscapes are human scale and link public transport to local destinations. Mixed-use neighborhood spaces promote dynamic living, working, socializing and recreation. The water sensitive detailing improves air quality and microclimate making the city attractive and comfortable.

德国戴水道设计公司受到委托，为台北市创建一处新的城市愿景。湿润的亚热带气候造成台风多发，洪水已成为城市之中一种常见现象，而当地的基隆河已严重退化。近些年来在强台风的作用下，2000mm的年降雨量被增强，在几天之内便可翻倍。在人口密集的城市之中，重黏土土壤将暴雨挑战更加复杂化。旨在拯救濒临灭绝的本地白鲟的项目也成为了创建一个新鲜、宜居的公共环境项目的契机，从本质上将雨水作为当地宝贵的一项城市资源。

连接地下蓄留渠的行道树成为了水敏城市设计的亮点，为附近的基隆河提供洁净水源。中央公园为城市带来一股清新的空气，城市湖泊还可以发挥收集、平衡台风暴雨的功用。50%的降雨能够实现就地管理，此洪水策略可以承受100年一遇的洪水侵袭。步行街道以人性化尺度进行设计，连接公共交通到当地各处。综合使用的社区空间促进了充满活力的生活、工作、社交和娱乐活动。水敏设计细节的应用改善了空气质量和城市小气候，将城市变得富有吸引力和宜居。

1 New mix development in the center of Taipei
台北市中心区域新的复合开发项目

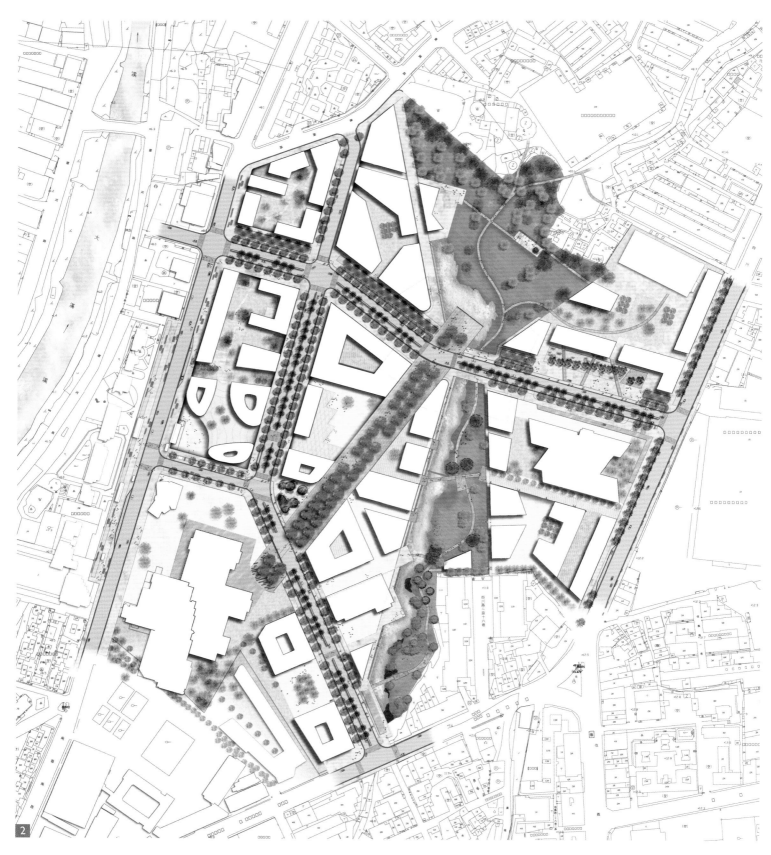

2 Masterplan with the central green lung
中央绿肺概念的规划方案
3 Water infrastructure for sustainable stormwater
便于可持续雨水管理的水基础设施

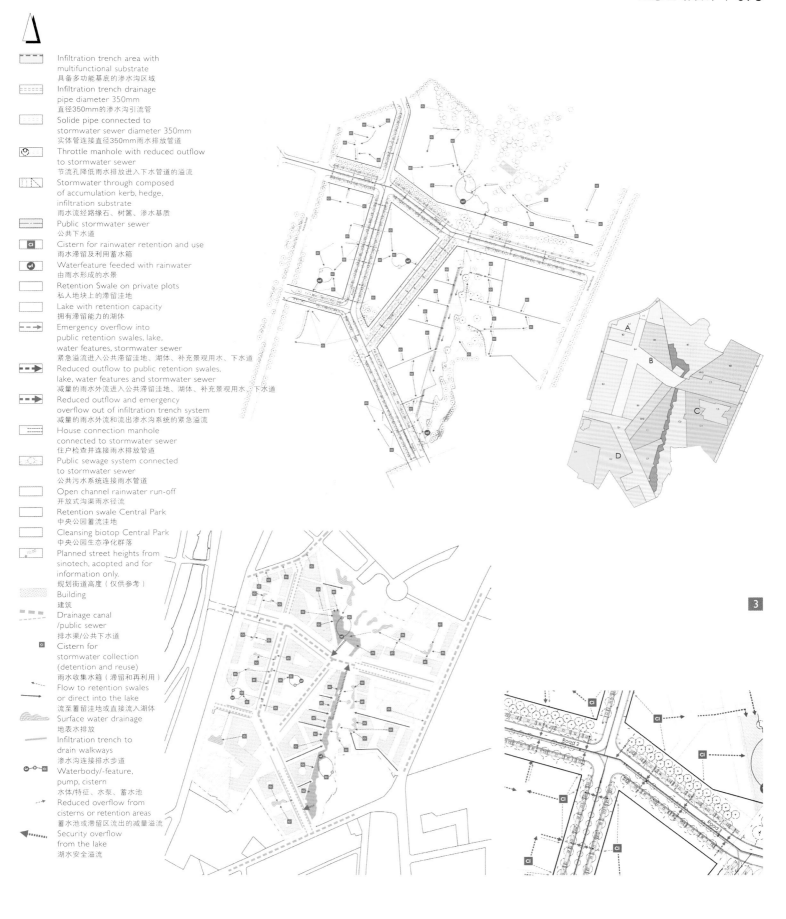

Cleansing Biotope Type I: Unsaturated Filter Material
生态净化群落类型一：不饱和过滤构造

- Well point-multiple locations 多点的井口位置
- Clay material barrier 黏土材料隔层
- Note: Plants selected for cleansing biotope to remain consistent and to extend along shoreline into and including lake
  生态净化群落选用植物与湖岸及湖中植物保持一致
- Height of water level 500~1,000mm below top of ground level
  水位高度低于地面最高处500–1,000mm
- Filter media 过滤介质
- Gravel subbase 碎石底基层
- Perforated pipe to lake 湖体穿管
- Water level Lake 湖体水位
- Drainage layer I-liner below lake bottom 排水层I——位于湖底下层

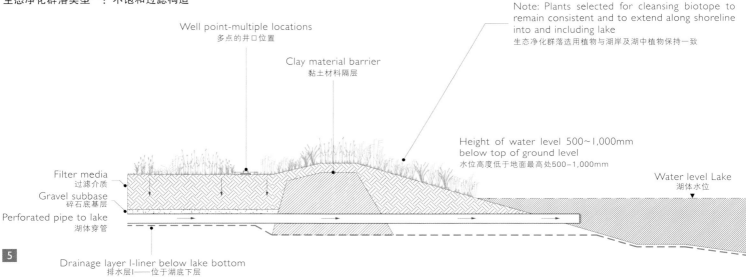

Cleansing biotope type II: Saturated filter material:
生态净化群落类型二：饱和过滤构造

- Pumped circulation volume 泵抽水量
- Well point-two locations 两处井口位置
- Barrier to be decided 隔板
- Note: cleansing biotope II allows water to remain at surface level. Upon reaching well points, water to seep through filter media, into perforated pipe, and direct into lake.
  2号生态净化群落保持表面水位，最高到达井口位置，水体渗透通过过滤介质，由湖体穿管直接流入湖体
- Filter media 过滤介质
- Gravel subbase 碎石底基层
- Perforated pipe to lake 湖体穿管
- Water level Lake 湖体水位
- Drainage layer 1 – liner below lake bottom 排水层1-位于湖底下层
- Drainage layer 2 – before barrier edge 排水层2-位于隔板边缘之前

4 Birdeye view of the north park
   北部公园鸟瞰图
5 Unsaturated treatment filter
   不饱和处理过滤
6 Saturatted water treatment filter
   饱和水处理过滤
7 The lake with the urban accessable edge
   湖体连接城市开放区域
8 In park integrated cleansing biotope
   整合生态净化群落的公园设计

B03 | Ecocity District | 076

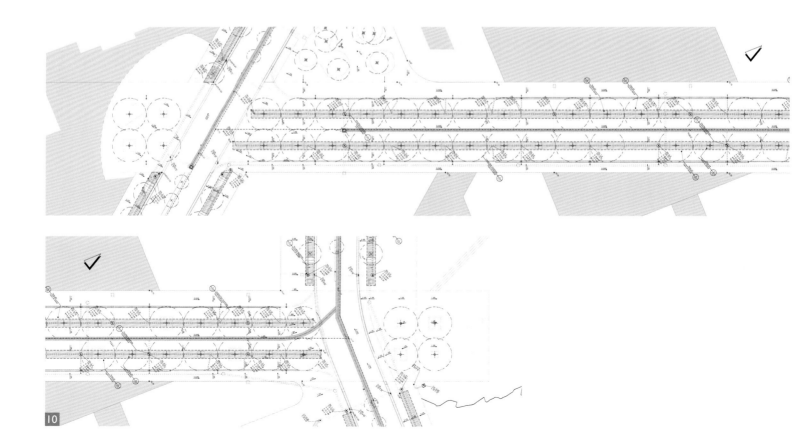

生态区域设计 | 077

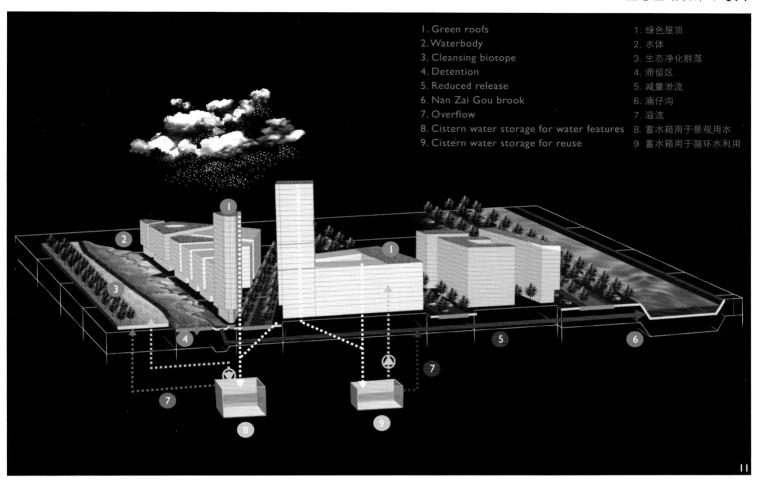

1. Green roofs	1. 绿色屋顶
2. Waterbody	2. 水体
3. Cleansing biotope	3. 生态净化群落
4. Detention	4. 滞留区
5. Reduced release	5. 减量泄流
6. Nan Zai Gou brook	6. 浦仔沟
7. Overflow	7. 溢流
8. Cistern water storage for water features	8. 蓄水箱用于景观用水
9. Cistern water storage for reuse	9. 蓄水箱用于循环水利用

9 East west boulevard
    东西大道
10 Stormwater infrastructure in streets
    街道之中的雨水基础设施
11 Model of the system
    系统模型
12 Water mangement for monsoon rain event
    季风性降雨雨水管理

Rd. #7 plan south-west scale 1:250:
西南部7号公路规划 比例尺1:250

Infiltration trench area with multifunctional substrate	具备多功能基底的渗水沟区域
Infiltration trench drainage pipe diameter 350mm	直径350mm的渗水沟排水管
Infiltration trench solide pipe diameter 350mm	直径350mm的渗水沟实壁管
Inspection manhole infiltration trench diameter 500mm	直径500mm的渗水沟检查井
Solide pipe connected to stormwater sewer diameter 350mm	直径350mm连接下水管道的实壁管
Throttle manhole with reduced outflow to stormwater sewer	节流孔降低雨水排放进入下水管道的出流
Stormwater through composed of accumulation kerb, hedge, infiltration substrate	雨水流经路缘石、树篱、渗水基质
Infiltration trench elongated after taipower construction	台电建设后延长的渗水沟
Gully open channel dimensions 500×500mm	500×500mm尺寸开放式水渠
Drainage channel boulevard typ 1 nw 150mm with gully dimensions 500×210 mm	西北部一号林荫道150mm排水渠和500×210mm水沟
Drainage channel boulevard typ 2 nw 400mm with gully dimensions 500×500 mm	西北部二号林荫道400mm排水渠和500×500mm水沟
Public sewage system with inspection manhole	附有检查井的公共下水系统
House connection manhole connected to stormwater sewer	连接下水道的房屋连接井
Three pits road	三坑道路
Three pits boulevard	三坑道林荫道
Gully road channel with upstand kerb stone (planning by sinotech)	带有直立路缘石的道路排水沟
Upstand kerb stone at pedestrian crossing	在步行路交叉口处的路缘石
Connection pipe	连接管
CH 9.55 Cover height manhole	井盖顶层高度
IL ID 9.55 Invert level pipe from infiltration trench	渗水沟管底高程
IL SS 9.55 Invert level pipe to stormwater sewer	排向雨水管网的管道管底高程
IL G 9.55 Invert level pipe from overflow gully	溢流沟管底高程
BH 9.55 Bottom height manhole	井盖底层高度
BH 8.10 Bottom height stormwater sewer	下水道底层高度
CH 8.84 / BH 3.00 Cover and bottom height sewage system	下水系统顶层和底层高度
10.00 Planned street heights from sinotech	规划的街道高度

JTC Clean Tech Park, Singapore
新加坡JTC清洁科技园

Location Singapore
Client Jurong Town Council
Engineer Surbana International Consultants Pte. Ltd.
Area 50ha
Design Time 2009-2011
Construction Time 2011-2012
Awards BCA Greenmark Platinum for New Parks, 2011

项目地址 新加坡
项目委托 裕廊镇管理局
工程合作 盛邦国际咨询有限公司
项目面积 50公顷
设计时间 2009年~2011年
建成时间 2011年~2012年
所获奖项 2011年BCA新建公园类绿标"铂金"奖

1　Future goal is rainforest
　　未来目标位创建雨林式园区
2　Public facility building
　　公共设备楼
3　Swales are integrated
　　洼地被整合入设计
4　Central building in the swamp area
　　沼泽区域的中心大楼

Envisioned to be the first business park set in a tropical rainforest, the JTC Clean Tech Park plays an important role in spear-heading Singapore's efforts to be a leader in the global thrust towards sustainability. It spans over a 50 hectare site, with a 5 hectares green core situated at its heart. Designed to be its lungs, the green core provides a place of respite for both the human inhabitants as well as the ecology of the site.

Building clusters are organized to have an urban front on one side, and a forest front on the other. Existing eco-habitats, including grassland, woodlands and peats, are retained as much as possible. Existing wildlife species were documented and a natural wildlife corridor connecting the site to the larger surrounding environment is enhanced through additional planting that provides food and habitat for them.

The natural topography is retained and natural water elements are implemented to support the existing hydrological flow of the site – bio-swales purify rainwater while channeling it from the roadside drains into the central core, where it will be retained in a swamp and circulated through a cleansing biotope for further purification, before being reused for toilet-flushing.

A deep respect for nature underlies this holistic design approach. It reflects mankind's desire of not only coexisting sustainably with nature, but engaging with and learning from it.

JTC清洁科技园区被构想为设置于热带雨林地区的首个商业园区，在发挥新加坡作为推动全球可持续性领导者角色方面发挥了重要的作用。园区占地面积50公顷，一个5公顷的绿色核心坐落在它的中心区域。作为设计的园区绿肺，此绿色核心既为人类居住者同时也为该场地内的生物提供了一个栖息场所。

建筑群的一侧与城市相接，而另外一侧则朝向森林。现有的生态栖息地，包括草地、林地和泥炭沼泽区都被尽可能地保留。现存的野生动物物种被记录，通过增加的植被种植，为野生动植物提供食物和栖息地，自然野生动物廊道功能被增强，将场地与更大范围的周边环境相连。

自然地形被保留，天然的水元素被应用，以支持现有的场地水文流动 – 生态洼地净化雨水，同时引导雨水从路边的排水渠进入中心地带。在那里，雨水将被保留在沼泽湿地之中，并通过生态净化群落进行循环和进一步的净化处理，被重新利用成为厕所冲刷用水。

对大自然的深深敬意成为了这个具有历史意义的设计方案的基调。这不仅反映了人类与自然可持续共生的愿望，而且鼓励与大自然接触并向它学习。

B04 | Ecocity District | 080

Site and Development Plans
场地开发规划

Flora + Fauna
动植物群落设计

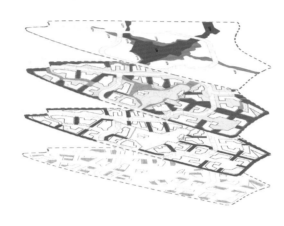

Art and Space Concept
艺术和空间概念

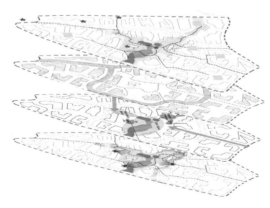

Stormwater Concept
雨洪管理概念

Landscape Concept
景观开发概念

5 Bioswale and composting station
  生态洼地和堆砂站
6 Local art work "Sculpted Maze"
  场地艺术品"雕刻迷宫"
7 Summit overlook
  瞭望台
8 Ravine stream
  溪流渠道

Kallang Riverside, Singapore
新加坡加冷河滨区域规划

Location Singapore
Client Urban Redevelopment Authority (URA)
Partners CPG Corporation Ptd Ltd,
Ernst & Young
Area 23.5ha
Design Time 2012~2013
项目地址 新加坡
项目委托 城市重建局
合作机构 CPG Corporation Ptd Ltd,
Ernst & Young
项目面积 23.5公顷
设计时间 2012年~2013年（概念设计阶段）

In Singapore's 2008 Master Plan, the Urban Redevelopment Authority identified Kallang Riverside as a growth area to be developed over the next 10 to 15 years. Sited along the waterfront and in close proximity to the city, the 23.5ha area will be a prime residential and hotel urban village developed to be the first water sensitive urban design (WSUD) precinct in the region with a high expectation towards a sustainable city.

WSUD does not equate to utilitarian design, water is celebrated in iconic features – a natural swimming pool, cleansing biotopes and art features which double as water monitoring stations. Generous park space, continuous promenade along the waterfront, "fenceless" residential developments, dynamic multi-functional areas, pedestrian oriented zones and site specific WSUD architecture come together in the spatial and environmental planning of the precinct, making a bold statement on the future of urban planning in land scarce Singapore: quality living spaces for people in environmentally responsive ways.

在2008年新加坡的总体规划中，城市重建局就已经将加冷河畔确定作为未来10~15年之中的一处开发区域。沿着海滨选址、且邻近城市，此面积23.5公顷的区域将被开发成为一处优质的拥有居住区和酒店的城市社区，成为该范围内拥有高标准、可持续城市展望的第一处水敏感城市设计区域。

水敏感城市设计有别于功利性设计，在它之中水体以一种特征标志的形式出现——自然式的泳池、生态净化系统和水景艺术形式。除却它们自身的功能性，还可以兼作为水质监测站。宽大的公园空间、沿水岸连续的步行道路、无围墙的居住区开发、富于变化的多功能区域、行人使用区域，以及具有场地独特性、符合水敏感城市设计理念的建筑形式，在该区域的空间和环境规划之中有机结合，为土地稀少的新加坡未来城市规划做出了一个大胆的声明：以环境敏感的方式为人们创造有品质的生活空间。

1 Sustainable development at the Kallang River
沿加冷河的可持续开发

B05 | Ecocity District | 084

2   Existing site at the river
    河滨既存场地
3   Water management plan for buildings
    楼宇水管理规划
4   Our vision for Kallang Riverside is to create a healing city by regenerating the connection between people, water and the environment through water sensitive urban design. This can be achieved by turning water management into the main driver of urban form.
    我们对于加冷河滨区域规划的愿景是通过水敏城市设计重新建立起人、水、环境之间的连接，创建一座具备治愈功能的城市。这可以通过将水资源管理作为城市形态的主要驱动力来实现。

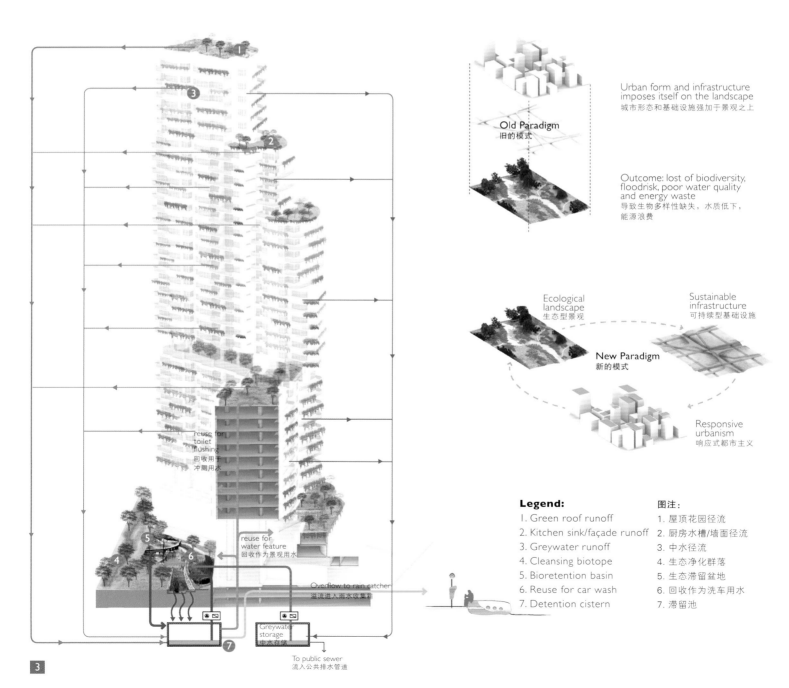

Urban form and infrastructure imposes itself on the landscape
城市形态和基础设施强加于景观之上

Old Paradigm
旧的模式

Outcome: lost of biodiversity, floodrisk, poor water quality and energy waste
导致生物多样性缺失，水质低下，能源浪费

Ecological landscape
生态型景观

Sustainable infrastructure
可持续型基础设施

New Paradigm
新的模式

Responsive urbanism
响应式都市主义

Legend:
1. Green roof runoff
2. Kitchen sink/façade runoff
3. Greywater runoff
4. Cleansing biotope
5. Bioretention basin
6. Reuse for car wash
7. Detention cistern

图注：
1. 屋顶花园径流
2. 厨房水槽/墙面径流
3. 中水径流
4. 生态净化群落
5. 生态滞留盆地
6. 回收作为洗车用水
7. 滞留池

生态区域设计 | 085

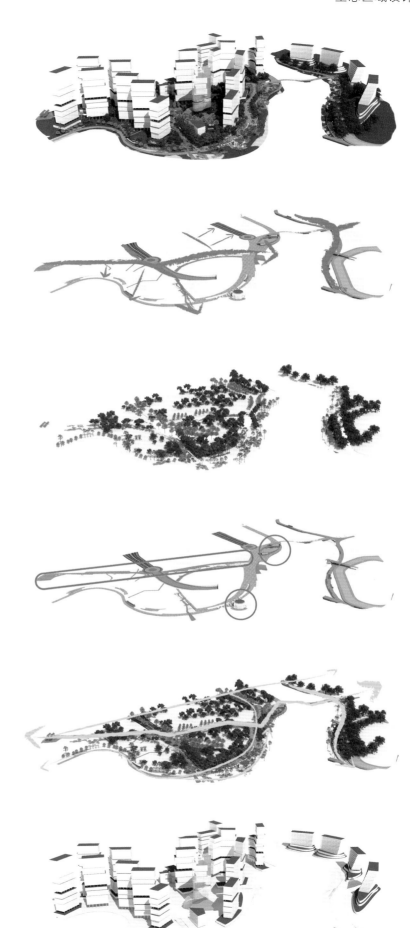

 Hydrology
水文

 Landscape
景观

 Culture
文化

 Connectivity
连通性

 Built Form
建筑形体

5,8 Roads and pedestrian areas drain in parks
公园内的道路和步行区域排水

6,9 Green corridor connects to the river
绿廊连接河流

7,10 Integrated treatment stripes along the promenade
与步行道整合的条状植被处理

在未来，对于河流内部动态过程的深刻理解应当成为所有可持续的跨学科综合开发项目的出发点，以便实现河流修复设计之中遇到的多种需求和挑战之间更好的整合。在实现这一目标时，三个方面至关重要：即分别为水体、动植物和人类活动提供更多空间。这种考虑展示出了一种处理往往看似完全不可兼容的需求关系的全新的协同可能性。当然，这也提出了一种挑战，即不可能在单独的一门学科之内解决问题。鉴于此，基于水利、生态、城市规划和景观设计所共有的制约性所进行的观察和决定，便显得很有必要。不同的专业之间需要开发制定出一种共通的语言，并建立跨学科工作结构。项目的筹备进行不应只由一个学科主导，其他学科仅在稍后的阶段介入而被剥夺了在基本概念规划决策阶段参与的机会。跨学科的组成以及规划团队之间自项目初始阶段的平等合作，对于项目最终质量发挥着至关重要的影响。

对于各学科的关键挑战在于未来的设计如何能够适应水体不断变化的节奏和动态能量。不要仅仅把河流想象成为阳光明媚天气时处于正常水位时的状态，也要考虑到干旱和洪水暴发的极端事件。城市河流空间是展现生态的、面向过程的设计和流域管理的极佳场地，在这里，自然变化规律、市政工程系统以及设计的景观进行叠加，随着气候等不断变化的条件状况持续地彼此影响并进行重塑。每个项目都展示了自己独特的挑战，每一处水体都反映不同的状态，每一处水体都存在不同的可用空间。面向过程的设计是指面对各种选择、后续措施以及自然发展对策的思考和规划。对于许多地方管理机构和规划师而言，这一"进化式"的设计方式是全新的；然而，它却对我们城市河流空间的未来具有极大的重要性。通过开发一个可持续的、跨学科的规划和设计文化，我们将能够重新恢复以城市河流为基本、构建功能强大城市空间的能力。

Bishan-Ang Mo Kio Park and Kallang River, Singapore
新加坡碧山宏茂桥公园和加冷河修复

Location Singapore
Client Public Utilities Board (PUB)
Partner CH2M HILL
Area 62ha
Design Time 2007-2011
Construction 2012
Awards WAF Landscape of the Year Award 2012, Excellence on the Waterfront Honor Award 2012

项目地址 新加坡
项目委托 新加坡公共事务局
合作机构 CH2M HILL
项目面积 62公顷
设计时间 2007年–2011年
建成时间 2012年
所获奖项 2012年世界建筑节年度最佳景观设计奖；2012年新加坡水滨设计优秀奖

1 River as a new natural environment
  河流成为了一处新的自然环境
2 62ha park after redesign
  重新设计后的62公顷公园
3 During heavy rain high volume of storage
  暴雨之中大量的雨水储存能力

Bishan-Ang Mo Kio Park is one of Singapore's most popular parks in the heartlands of Singapore. As part of a much-needed park upgrade and plans to improve the capacity of the Kallang channel along the edge of the park, works were carried out simultaneously to transform the utilitarian concrete channel into a naturalised river, creating new spaces for the community to enjoy. At Bishan-Ang Mo Kio Park, a 2.7 km long straight concrete drainage channel has been restored into a sinuous, natural river 3.2 km long, that meanders through the park. Sixty-two hectares of park space has been tastefully redesigned to accommodate the dynamic process of a river system which includes fluctuating water levels, while providing maximum benefit for park users. Three playgrounds, restaurants, a new look out point constructed using the recycled walls of the old concrete channel, and plenty of open green spaces complement the natural wonder of an ecologically restored river in the heartlands of the city. This is a place to take your shoes off, and get closer to water and nature!

碧山宏茂桥公园作为新加坡市中心区域最受欢迎的公园之一，急需规划升级。与此同时进行的还有公园沿线的加冷河水渠修复计划，将单一功用性的混凝土结构河道转变为自然式河流，以创造社区居民能够充分享用的新型城市空间。在碧山宏茂桥公园，长2.7千米的笔直混凝土排水渠已经恢复为长3.2千米的弯曲、自然式河流，蜿蜒穿过公园。62公顷的公园空间被重新进行设计，以适应包括水位波动等的河流系统的动态过程，为使用者提供最大化收益。使用原有混凝土水渠改造而来的回收石材，建造了三个游乐场、一处餐厅以及全新的瞭望台。大量的绿色开放空间也为城市中心区生态修复河流形成的自然奇观提供了有益的补充。这里是一处可以脱掉鞋子，与水和自然亲近的地方！

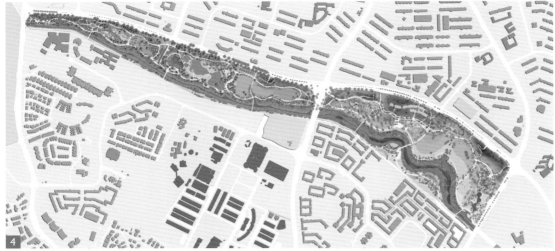

4 The new river connects city
   全新的河流连接着城市
5 Park and river are merging
   公园与河流相互融合
6 Bioengineering stabalize riverbanks
   生态工法加固河岸

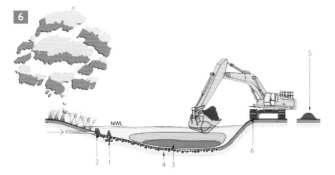

Sedimentation Basin:
1. Energy dissipation
2. Inlet structure
3. Sediment deposit
4. Hard rock bottom or defined sand layer
5. Area for sediment de-watering
6. Access zone for sediment removal

沉积盆地：
1. 能量损耗
2. 进气结构
3. 沉积物堆积
4. 坚硬的岩石底部或定义的沙层
5. 沉积物脱水区
6. 沉积物移除工作区域

Section:
1. Gabions with brush layers
2. Brush layer stone-filled wire-mesh
3. Gabion basket (100×100×50)
4. Selected fill material
5. Gravel (sub-grade course)
6. Assumed mean water lever
7. Log cribwall
8. Brush layer and/or rooted plants
9. Wood cribwall built from timber logs
10. Bed substrate
11. Undisturbed existing soil

剖面图：
1. 附有植被冲刷层的石笼
2. 植被冲刷层 填石丝网
3. 石笼篮（100×100×50）
4. 选择填充材料
5. 砾石层（路基）
6. 假设平均水位
7. 木质框架挡土墙
8. 植被冲刷层和/或生根固土植物
9. 木材建造的木质框架挡土墙
10. 河床基底
11. 未受干扰的既存土壤

7 New park as a place for people and streams
新建公园为人们和溪流提供了共存的空间
8,9 Biodiversity is coming back
生物多样性的回归
10 Students showing water testing results from the river to the Prime Minister
学生们向总理展示河流水质测试结果

河流修复 | 097

10

11

12

11 New pedestrian bridge
   新建步道桥
12 Bubble Playground
   气泡游乐场
13 Relaxing at the recycling hill
   在采用回收材料建造的山丘上休闲放松
14 People get close to the water
   人们与水接近

Daylighting of Alna River, Holalloka, Oslo, Norway
挪威奥斯陆海伦拉卡，埃纳河的日光

Location Oslo, Norway
Client City of Oslo
Partner 13.3 Landscape Architects
Area 2ha
Completion Time 2007
项目地址 挪威 奥斯陆
项目委托 奥斯陆市政府
合作机构 13.3 Landscape Architects
项目面积 2公顷
建成时间 2007年

The "daylighting" of a section of Alna is combined with stormwater management and an urban park with informal recreation areas. The project is a pilot project for the large-scale restoration of the river Alna. The Alna is released from its old concrete pipe and restored to its natural form of soft planted banks and pond areas. This re-naturalization not only provides habitat creation and a distinctive green character to an urban park, it ensures important river management functions.

High quality and artistic design integrates the restored natural qualities of the river into the urban fabric. New pathways link the river into local circulation and provide new recreation possibilities. The quality of the water is good enough to swim in! A bathing pond with a beach is the central focus of the park. A wooden deck is a comfortable sunning area with views to cascade elements which double as a stepping stones and as sized detention outlets. The technical aspects of water management are integrated with humour and flair into the attractive and relaxed atmosphere of the park. The stormwater management of the adjacent industrial buildings is combined into the river system. Rainwater is collected from the roofs and conveyed via surface details into a planted cleansing biotope area. Here the rainwater is cleansed through a multi-layered substrate and held back for slow release.

埃纳河的"日光"部分，结合了雨水管理与非正式休闲城市公园的功能。该项目是大型的埃纳河恢复项目的一个试点工程。从老旧混凝土管道之中解脱出来，埃纳河被修复成为拥有软质植被河岸和池塘区域的自然形式。这样的"再自然化"不仅提供了栖息地和具备独特绿色特征的城市公园，同时还确保了重要的河流管理功能的发挥。

高质量和艺术化的设计将修复河流的自然本色整合入城市整体结构当中。新设计路径连接河流与当地道路交通，提供新的娱乐场地。河流水质良好，甚至可以进入游泳！拥有岸滩的游泳池成为了公园的中心焦点。一个木制甲板是一个舒适的日光浴场地，可以观赏同时作为踏步石以及大大小小排水口的水瀑元素。水管理的工程技艺以幽默和充满才华的方式被展示出来，融入公园具有吸引力、轻松愉悦的氛围之中。相邻工业建筑的雨水管理与河流系统有效结合。从建筑屋顶进行雨水收集，通过地表的细节处理被引导进入一片植物生态净化群落。在这里，雨水通过多层土壤基质的过滤得到净化，并放缓脚步缓慢流淌。

1,2 Alna opening creates swimming pond
埃纳河口区域游泳池

3   Granit slabs create an overflow
    花岗岩铺装形成了一处溢流景观
4   Woodendecks and swimming pond
    木质甲板和游泳池
5   Cross path and flow control in one
    通行道路和水流控制合二为一
6   Masterplan of Höllalökka
    海伦拉卡平面规划图
7   Cleansing biotopes adjacent to the stream
    紧挨溪流的生态净化群落

Site Plan:
总平面图
1. Natural area with rare vegetation
   拥有珍稀植被的自然区域
2. Lawn
   草坪
3. Sedimentation lake/pool
   沉淀湖/池
4. Viewpoint
   观景点
5. Swimming pool with stormwater detention
   雨洪滞留功能的游泳池
6. Path with gravel
   碎石路径
7. Wooden deck
   木甲板
8. Water stairs/cascade
   水阶梯/水帘
9. Wetland with stormwater detention
   雨洪滞留湿地
10. Cleansing biotope
    生态净化群落
11. Stormwater catchment area
    雨洪汇集区
12. Open stormwater channel
    开放式雨水渠

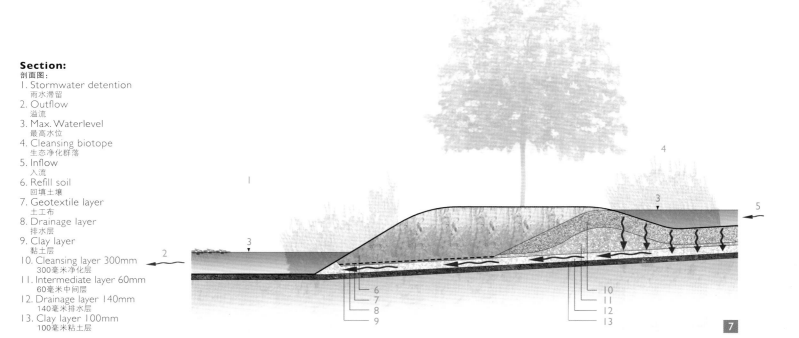

Section:
剖面图：
1. Stormwater detention
   雨水滞留
2. Outflow
   溢流
3. Max. Waterlevel
   最高水位
4. Cleansing biotope
   生态净化群落
5. Inflow
   入流
6. Refill soil
   回填土壤
7. Geotextile layer
   土工布
8. Drainage layer
   排水层
9. Clay layer
   黏土层
10. Cleansing layer 300mm
    300毫米净化层
11. Intermediate layer 60mm
    60毫米中间层
12. Drainage layer 140mm
    140毫米排水层
13. Clay layer 100mm
    100毫米粘土层

C03 | River Restoration 河流修复

Restoration of River Volme, Hagen, Germany
德国哈根市沃尔姆河修复

Location Hagen, Germany
Client City of Hagen
Partner PASD Architects
Area 1.2ha, River 800m
Design Time 2001-2006
Completion Time 2006
项目地址 德国 哈根
项目委托 哈根市政府
合作机构 PASD Architects
项目面积 1.2公顷，河流长800米
设计时间 2001年-2006年
建成时间 2006年

1,2 The river volme back to the awareness
重回公众视线的沃尔姆河

Urban and riverside planning is with the goal to increase the overall attractiveness of the city of Hagen. Previously, the Volme existed as a channeled, lifeless waterway without access. The first part of this long-term redevelopment project is the restoration of the embankment of the river Volme between the weir "Kaufmannsschule" and the bridge "Marktbrücke".

In the entrance hall of the new city hall building a color ful water and light band symbolizes the intimate connection between the city and the river. The glass sculpture adds light to the entrance and water flows down the irregularly shaped surface. The movement of the water, the sound and the refraction of the light create a holistic artistic statement. "Threads" of light and colored glass laid into the floor of the city hall symbolize the aquatic landscape around the city of Hagen.

On the outside of the building cascading water falls down the artistically designed staircase leading to a patio that opens to a view of the river Volme, creating a strong connection between the river and the residents of Hagen. As long as the river isn't flooded, one can reach the Kaufmannsschule by using a similarly designed staircase immediately next to the river.

城市和河岸规划的目标是提升哈根城市的整体吸引力。在此之前，沃尔姆仅仅作为一条河渠，毫无生气、且无法进入。此长期重建项目的第一部分，即为恢复位于康夫曼舒大坝与马卡布鲁克大桥之间的沃尔姆河路堤。

在新市政厅建筑的入口大厅，丰富多彩的水、光带象征着城市与河流的密切联络。玻璃雕塑为入口空间增添亮色，水体沿着不规则形状的表面向下流动。水的运动、声响以及光的折射，创建了一个整体的艺术化表达。光线以及铺设于市政厅地板内的彩色玻璃象征着水景观环绕哈根城市的周围。

在建筑外部，跌落的流水顺着艺术化设计的阶梯流淌而下，进入能够欣赏到沃尔姆河景色的露天庭院，在河流与哈根市居民之间建立起强大的连接。当河流未涨水之时，人们可以通过一个拥有相似设计的、紧邻河流的台阶到达康夫曼舒大坝。

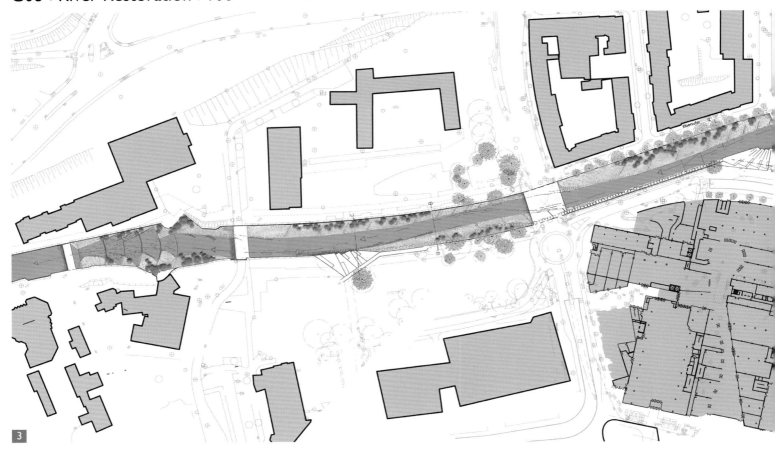

3 Masterplan for 1 km river
   1000米河域规划
4-7 River as a new experience for people
   治理过的河流区域成为了一处新的公众体验场所
8,9 Weires are transformed into ramps
   堰坝被转换为坡路

河流修复 | 107

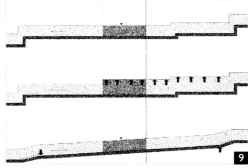

Rochor Canal, Singapore
新加坡梧槽运河修复

Location Singapore
Client Public Utilities Board (PUB)
Partner Surbana Engineering
Area About 1.2km long Rochor canal (serving catchment area of about 650ha)
Design Time 2010
Construction Time 2011-2013
Expected Completion Time 2014
项目地址 新加坡
项目委托 新加坡公共事务局
合作机构 盛邦国际咨询有限公司
项目面积 1.2千米运河，服务水域面积650公顷
设计时间 2010年
施工时间 2011–2013年
预计建成时间 2014年

Rochor River is an urban stormwater canal which flows from Bukit Timah all the way down to Marina Barrage. The canal flows mostly through a dense urban fabric and flows along stretches of heavy traffic roads. The new role of the Canal is to create a 'tie' connecting the segregated enclaves together, and give a stronger sense of the Marina, the City and its water edge. Accompanied with parallel pedestrian boulevards from the surroundings - Rochor Promenade, a waterscape which combines both green, blue and orange (human) elements - will soon be bursting activities and life. At the same time, the green belt offers a green corridor for fauna along the canal, enabling residents an opportunity to appreciate the wildlife at their doorstep. The design of the new Rochor Promenade is intended to bring people closer to the water edge. A layering system for the different city infrastructures of green, vehicular, pedestrian, culture, etc is adopted so that the design can react flexibly to the current conditions and programs while still being united as a whole.

梧槽河是一条城市雨水管道，从武吉知马一路延伸至滨海堤坝。运河流经人口密集的城市区域，并沿主要交通干道流淌而下。运河的新角色是创造一个连接分散地块的纽带，并赋予海滨区域、城市和水岸更强烈的场地感。与周围环境之中的行人大道——梧槽长廊平行，一个结合了植被、水体和人的元素的水景观将会很快激活该地区的各项活动和人们的生活。同时，沿运河设置的绿带为动物提供了一条绿色长廊，能够让居民在自家门口不远处即可观察到野生动物的活动。新的梧槽长廊的设计是为了让人们更接近水边。植被、行车、步行、文化等不同城市基础设施分层系统被制定，因此该设计可以在作为一个有机整体的同时，仍然能够灵活应对现时的状况和项目需要。

1,2 Rochor Canal as a new life line in the city
梧槽运河成为了城市当中全新的生活迹线

C04 | River Restoration | 110

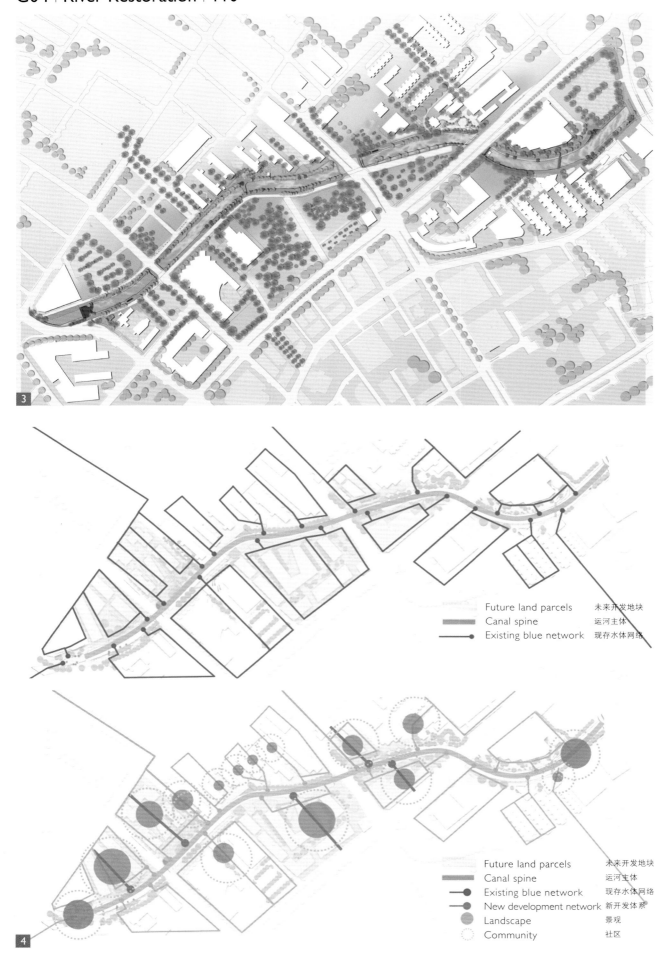

3 Attractive recreation waterfront
  有吸引力的休闲水滨
4 Stormwater management combined
  综合雨水管理
5 Investigate different usergroups
  对于不同使用群体的调查
6-8 Promenade for all needs
  满足多种需求的步道设计

The Edge   区域边缘

General pedestrian   普通行人

Residents   居民

Students   学生

Commuters   上班族

Business/workers   商业/工人

Joggers/ cyclists   跑步者/骑车人

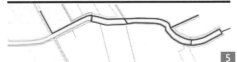

Feng Chan River Restoration, Zhangjiawo New Town, Tianjin, China
中国天津张家窝新镇，丰产河修复

Location Tianjin, China
Client Shanghai Wisepool Real Estate Co., Ltd.
Architect Theodor Schaller Schmitz Architekten
Area masterplan 180ha, 20ha (phase 1)
Design Time 2005 - 2008
Construction Time 2006 - 2008
项目地址 中国 天津
项目委托 上海国民实业有限公司
建筑设计 Theodor Schaller Schmitz建筑事务所
项目面积 总规划面积180公顷，一期工程20公顷
设计时间 2005~2008年
施工时间 2006~2008年

A water sensitive urban design concept creates a language of green street detailing. Rainwater is collected, cleansed through natural plant systems and stored in a central waterbody. Small pedestrian neighbourhoods are grouped together creating both privacy and community. This ecological infrastructure concept is vital to prevent brackish water pushing in from the sea.

The Feng Chan River has been restored from a 30-metre wide polluted irrigation canal back into a beautiful, clean, natural river. The banks are softly shaped and planted – a flourishing ecology has rebalanced the natural biology of the water making the Feng Chan River a community asset which is a focus point for leisure and socialising, as well as safeguarding water resources.

水敏感城市设计概念为绿色街道细部设计提供了新的形式。通过自然式植被体系，雨水被收集、净化，存储于中央水体。小型步行街区被组合在一起，创建私密空间、组成和谐社区。生态基础设施的概念能够有效地防止苦咸海水的涌入。

丰产河由一条30米宽、饱受污染的灌溉运河恢复成为美观、清洁、自然的河流。蜿蜒的软景河岸植被丰富，一个蓬勃发展的生态环境很好地平衡了水体的自然生物系统，将丰产河转变成为一种社区资产，一处休闲、社交的焦点，同时也对保护地方水资源发挥了一定的作用。

1 Clean riverspace for people of Zhjangiawo
  为张家窝居民设计建造的整洁河滨区域
2,3 Clean riverfront becomes social place especially for children's play.
  洁净的河滨区域成为了人们聚集的场所，孩子们尤其喜欢在这里嬉戏玩耍

4　River is now in positive image
　　清洁、美观的河流
5,6　Park along the river attracts people
　　沿河公园吸引了人们前来
7　A distinct concept of sealing was choosen
　　独特的不透水概念被加以应用
8　Stormwater overflow leads to the river
　　雨水溢流流向河流

河流修复 | 115

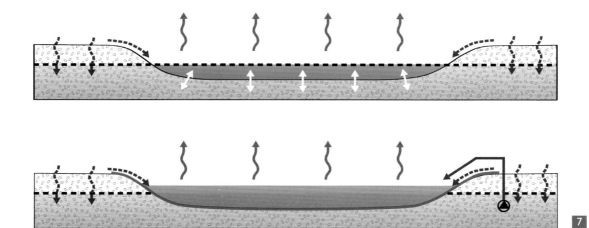

**Stormwater System
(Zhangjiawo New Town):**
张家窝新镇雨洪系统规划：

— Infiltration ditch
　渗水沟
— Infiltration ditch filled
　填充的渗水沟
— Open drainage gully
　开放式排水沟
■ Orchard retention area
　果树滞留区
⊙ Pump chamber
　泵室
◣ Overflow to river
　溢流至河流
▪▪▶ Emergency overflow to river up/downstream
　紧急溢流至河流上/下游

FENG CHAN
RIVER
丰产河

Cities Moving toward Sustainability
by Khoo Teng Chye

'Our waterways and reservoirs should do more than meet our water needs. They should also enhance our living environment and lifestyle. By linking up our water bodies and waterways, we will create new community spaces that are clean, pleasant and bustling with life and activities. We will integrate our water bodies with our parks and green spaces and turn Singapore into a "City of Gardens and Water".'*

More than half of the world's population is living in cities. Despite 200 years of hydrological engineering history, achieving water sustainability in our cities remains a major challenge. The issues of water for cities are not restricted to providing adequate water supply for residents, businesses and industries. It is vital to ensure that used water does not pollute rivers and that cities do not flood during rainstorms. And it is increasingly important that water infrastructure should not merely be functional, but collaborate with the greater water system and positively contribute to the social, cultural and recreational quality of urban living.

Singapore's integrated water resource management

Many modern cities are characterised by enormous growth and need large areas of land and resources to support their energy, mineral, food and water demands. Once these resources have been consumed, they are dumped as waste products into the surrounding landscape. This cycle is highly destructive; and a state of ecological collapse is evident in many cities in Asia, Europe and in fact across the world.

Uniquely, Singapore is a densely populated island city state; 4.8 million people live on an island of only 700 square kilometres. Although situated near the equator and hence receiving ample rain (2,400 millimetres per year), the country's small land area means that there is not enough space to collect and store all of the water it needs; nor does the island have any aquifers. Hence, ensuring water sustainability is a strategic, national challenge. Perhaps because space and resources are at such a high premium in Singapore, we have had to review our water system beginning in the 1970s, earlier than many other cities. Evolving from this, today Singapore employs a holistic approach towards managing its water needs. This involves diversifying its supply sources, managing demand through a water conservation strategy, and ensuring shared ownership of water resources by stakeholders (the people, public and private sectors). This approach has gained international recognition, especially for Singapore's efforts at large-scale water reclamation. Water of highest quality is reclaimed from waste water using advanced membrane technology, resulting in the end product called NEWater.

Equally significant is Singapore's approach to harvesting rainwater on a large scale. This has been done over several decades, through systematic land use, environmental management and the separation of the stormwater and sewer systems. Singapore now has 15 rainwater reservoirs. Half of the island serves as water catchment, and this will soon expand to two-

* From Singapore's Prime Minister Lee Hsien Loong's speech during the launch of the Active, Beautiful and Clean (ABC) Waters Public Exhibition in February 2007

thirds of the island supported by the completion of the Marina Reservoir and two other reservoirs. Recognising our urban environment as a catchment is a significant step towards rebalancing management of the water system, and by default restores the importance of the relationship between individuals and the water they see, use, experience and enjoy. Marina Reservoir, located in the downtown area, was formed by erecting a barrage across the mouth of the Marina Channel that connected Marina Bay to the sea. As the bay was formed by reclaiming land to create Singapore's new downtown area, the new Marina Reservoir is located as a celebrated centrepiece and will be a vibrant body of common property water in the heart of the city.

Urban revitalisation – the ABC (Active, Beautiful and Clean) Waters Programme

The ABC Waters programme identifies the task of main taining, improving and extending this water resource as the opportunity not only to improve water quality but to be an urban driver revitalising the city and improving accessibility, aesthetics, recreational potential and ultimately forming a bond of responsibility between individual citizens and the nation's water. Its aim is to turn Singapore into a 'City of Gardens and Water'. Essentially, it will bring people closer to the water.

First, a comprehensive approach was taken to identify potential projects across the island and to implement them systematically over the next 10 to 15 years. Singapore was divided into three watersheds – Central, Eastern, and Western Singapore. Three different multidisciplinary consultants (one of which is Atelier Dreiseitl) were then engaged, each to develop plans for one of the watersheds. The proposals by these watershed managers took into account land use plans, development plans, demographics, hydrology as well as historical, cultural, educational and safety considerations.

From the outset, PUB recognised the importance of actively engaging the public, private and people sector partners in all aspects of the ABC Waters programme. As each catchment master plan was drawn up, key proposals were presented to other public agencies, potential stakeholders and non-governmental organisations, as well as a panel of building and construction experts. The feedback from the various groups was used to further refine the master plans.

Singapore ABC Waters programme is a vision for tomorrow which is happening today. As a city we have provided a framework for an integrated planning process, where on each level multi-disciplinary teams collaborate creatively within themselves and with the various city departments and stakeholder groups involved. This is not just a good 'idea' but is being put into practice and can be seen in outstanding projects such as the river canal restoration in Bishan Park, designed by Atelier Dreiseitl.

迈步走向可持续化城市

邱登才

"我们的水道和水库应该不止于满足我们对水的需求,它们还应该在提升我们的生活环境和生活方式方面发挥作用。通过连接水体和水道,我们能够创建清洁、舒适、充满热闹生活和活动的新型社区空间。我们将水体结构与公园和绿地空间整合在一起,使新加坡成为一个'花园和水景城市'。"*

全世界有超过一半的人口居住在城市。尽管人类拥有200年的城市水文工程历史,实现城市之中水资源的可持续发展依然面临很大挑战。解决城市水问题不仅仅局限于为居民、商业和工业提供充足的水源,更为重要的还在于确保废水不会污染河流以及保护城市免受洪涝灾害侵袭。近些年来愈发深刻地认识到,水基础设施不应该仅仅具备功能性,而应与更多的水系统结合,对提高城市生活的社会、文化以及休闲活动品质做出积极贡献。

新加坡综合水资源管理

许多现代城市正经历着快速的发展和增长,需要大面积的土地和资源支持其能源、矿产、食品和水需求。一旦这些资源被消耗掉,它们会被作为废弃产品丢弃到周围环境之中。这个过程极具破坏性,亚洲、欧洲以及遍布世界各地许多城市的生态崩溃的状态即为其最充分的证据。

特别的是,新加坡是一个人口密集的海岛城市国家,480万人居住在面积只有700平方千米的一个小岛上。虽然位于赤道附近,并因此获得充足的降水(2400毫米/年),但它狭小的土地面积并不足以收集和储存所需要的全部用水,岛上也没有任何的蓄水层。因此,确保水资源的可持续性成为了一个战略性的国家挑战。

也许是因为空间和资源在新加坡是如此的昂贵,在20世纪70年代初我们不得不早于其他许多城市,开始检讨我们水系统的合理性。因此,今日的新加坡拥有对其所需水资源管理的整体性策略,其中包括供应来源多元化,通过水资源保护战略进行需求管理,以及确保利益相关者(公众、公共和私营部门)共同拥有水资源。这种方式已经获得了国际性的认可,尤其是对于新加坡在大规模水资源回收方面所进行的努力,比如利用先进的膜技术回收废水成为高质量、被称为"新生水"的产品。

同样重要的还有新加坡所进行的大规模雨水收集。通过系统的土地利用、环境管理以及将雨水和污水处理系统的分开处理,雨水收集的做法在新加坡已经被执行几十年。新加坡国内目前有15个雨水储存库。一

*来自新加坡总理李显龙在2007年2月期间推出的"活力,美观,洁净"即ABC流域公众展览时的讲话。

半的岛屿面积可以作为集水区，并且伴随着滨海和其他两个水库的建成，集水区面积将很快扩展到三分之二的岛屿面积。认识到将我们的城市环境作为集水区，是平衡水系统管理的一个重要步骤。如此，很自然的即可恢复个人与他们所见、所用、所感、所体验的水资源之间的重要关系。滨海水库位于市中心区，通过建立拦河坝跨越滨海运河入海口、连接滨海湾入海组建而成。由于海湾是由回填土地建造并成为新加坡的新商业区，且全新的滨海水库正是位于其著名的核心部位，它的建成将成为城市的中心地带一处充满活力的国有水体资产。

城市复兴——"活力、美观、洁净（Active, Beautiful, Clean）"水域方案

ABC水域方案定义了在维护、改善和扩大水资源方面的主要任务。作为一个时机，它不仅有助于提升水质，而且可以成为城市重焕生机的驱动力：改善交通可达性、促进城市美观、增强休闲娱乐开发潜力，并最终形成公民个人与国家水体之间责任连接的纽带，达到将新加坡转变成为一个"花园和水景城市"的目标。从本质上讲，它有助于将人们与水的关系变得更为亲近。

首先，采用一个综合的方式确定横跨海岛的潜在项目，并在未来10~15年内进行系统地实施。新加坡被分为三个流域——中部、东部和西部流域。三个不同的多学科综合整治顾问团队（德国戴水道公司作为其中之一）被聘请，分别开展其中一个流域整治的规划工作。这些流域管理者们的提议之中综合考虑到了土地利用规划、国家发展计划、人口结构、水文以及历史、文化、教育和安全等方面。

从一开始，新加坡公用事业局（PUB）即认识到积极鼓励公共、私人和合作方人士参与、融入ABC水域规划项目方案之中的重要性。每一个流域总体规划的制定一经完成，其主要的规划即被提交给其他公共机构、潜在的利益相关者、非政府组织，以及建筑和施工专家们。从各组不同人群收集到的反馈意见则被用以进一步完善总体规划。

新加坡ABC水域方案是一个今天发生、面向未来的愿景。作为一个城市，我们为整合的规划过程提供了一个框架。在其中每一个层级的工作中，多学科背景的团队内部以及与所涉及各城市职能部门和利益相关者之间，都进行着富有创造性的通力合作。它不仅仅是一个很好的"想法"，而且已经被付诸实践，由德国戴水道公司设计的优秀项目——碧山宏茂桥公园之中的河道修复工程（加冷河修复）即是最好的代表。

Zollhallen Plaza, Freiburg, Germany
德国弗莱堡市扎哈伦广场

Location Freiburg, Germany
Client Aurelis Real Estate GmbH & Co.KG, City of Freiburg
Partner Misera Engineers
Area 5,600m²
Design Time 2009-2010
Completion Time 2011
项目地址 德国 弗莱堡市
项目委托 弗莱堡市奥瑞利亚斯房地产有限公司
合作机构 Misera Engineers
项目面积 5600平方米
设计时间 2009-2010年
建成时间 2011年

Zollhallen Plaza is new counterpart for the historic customs hall which was restored in 2009. The plaza is a fine example of water sensitive urban design, as it is disconnected from the sewer system. Beautiful planters provide infiltration points, and subsurface gravel trenches with innovative in-built filter medium reduce the hydraulic overload on the sewer system. Indented plaza areas create a surface flood zone. No rain water is fed to the sewer system, instead the groundwater table is recharged. The design plays with the historic past of the site which was a railyard. Timeless and multifunctional benches recalled break noses of railtracks, and old railtracks are inlaid into the paving. A bright grove of cherry trees provide the perfect amount of shade, while the infiltration planters with perennials and ornamental grasses give an attractive softness. 100% of the hardscape materials are high-quality demolition materials recycled from the old railyard. This makes sense not just from a resource management point of view, but harmonises the new clean modern design with the historic architecture of the customs hall.

扎哈伦广场的建成成为了2009年修复的历史悠久的海关大厦的鲜明对比。该广场完全脱离污水处理系统，成为了一个很好的水敏性城市设计的案例。美丽的种植池提供了渗透点，拥有创新式内置过滤基质的地下砂石沟渠减轻了污水处理系统的水压负载。缩进的广场区域创建了一个地表防洪区。雨水没有汇入地下污水处理系统，而是补给地下水位。该设计建造于曾经的铁路院落场地之内。富有年代感、多功能的座椅唤起了人们对于铁路轨道枕木的记忆，而旧的铁路轨道则被镶嵌于地面铺装之中。一片明快的樱花树林提供了充足的树荫，渗透式种植池之中育有多年生植物和观赏草类，形成鲜艳、柔美的花草景观。100%的硬质景观材料均来自旧的铁路院落场地回收利用的优质材料。因此，不论从资源管理的角度，还是建造材料的来源方面，这一新型、清洁的现代化广场设计与历史性海关大厦的建筑形式相互协调。

1　Plaza design respects the history
　　广场设计尊重场地历史
2　Plaza pavement and planting
　　广场铺装和树木种植

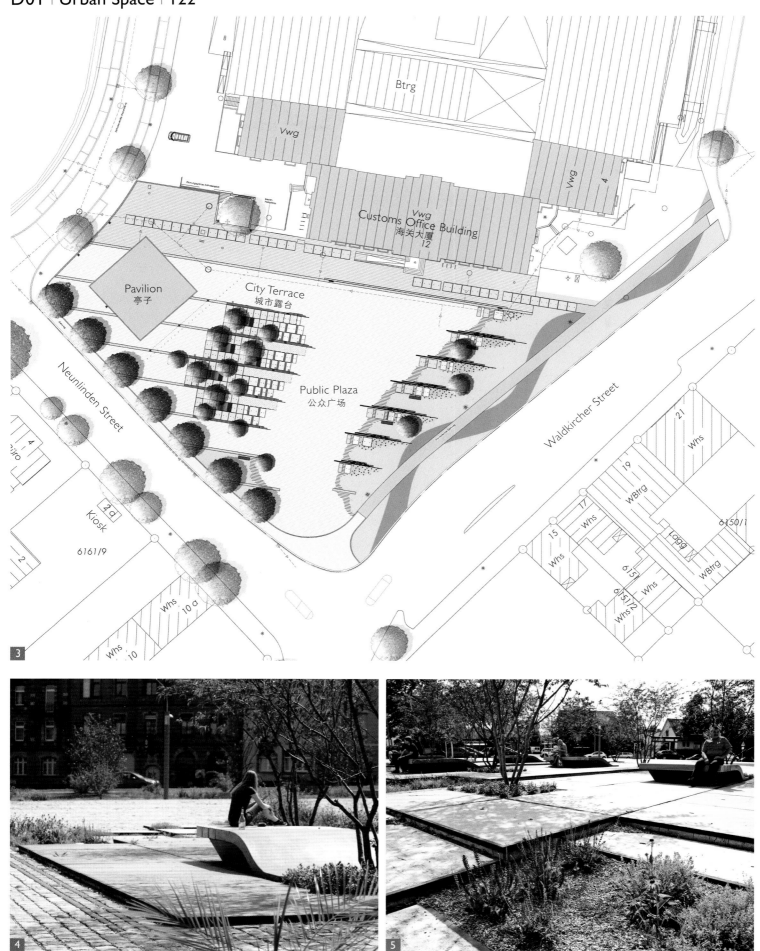

3 Design for the plaza
 广场设计
4,5 People like the casual furniture
 人们喜欢简易的休闲设施
6-9 Herbaceous planting strenghten the character
 草本种植形式强化了场地特征
10 Stormwater infiltration and storage underneath the plaza
 雨水渗透并存储于广场下方

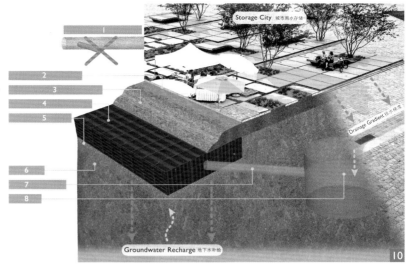

Cloudburst Plaza: 雨洪广场：
1. Disconnect from sewage　　1. 无需连接下水管道
2. Permeable paving　　　　　2. 透水铺装
3. Load-bearing substrate　　　3. 承重基层
4. Water storage boxes　　　　4. 蓄水箱
5. Filtration layer　　　　　　 5. 过滤层
6. Sub-soil　　　　　　　　　6. 底层土壤
7. Overflow pipe　　　　　　 7. 溢流管
8. Cistern　　　　　　　　　 8. 水槽

Watertraces, Hannoversch Münden, Germany
德国汉恩蒙登"水之印"广场

Location Hannoversch Münden
Client City of Hannoversch Münden
Area 1,000m²
Completion Time 2000
项目地址 汉恩蒙登
项目委托 汉恩蒙登市政府
项目面积 1000平方米
建成时间 2000年

With three rivers flowing nearby, water is an important component of the town of Hann. Münden. In addition to three rivers, the town also has three connected squares in the heart of the old quarter. When it was becoming nearly impossible to walk this area as a pedestrian, the people had the idea of redesigning and revitalizing the three squares as people places. A theme was quickly found: 'water-traces'- the paths followed by watercourses. Experts worked together with laymen in an open atmosphere. The Dreiseitl studio built its contribution to this system, which is fed mainly by rainwater, in the square between the church and the town hall. There are four terraced steps here, like a large folded carpet. Now people of all ages can go in search of the traces left by water. People can leave their own traces here as well. The nature of the flow is changed if people simply step into the pool. The flow pattern can also be changed by using the wave-making devices that are placed around the carpet. V-shaped glass-elements, about 5m high, hat sit on a steel plinth and are lit obliquely after dark.

三条河流流经汉恩蒙登市附近区域，因此水成为了当地一个重要的组成部分。除了三条河流之外，三个相互连通的广场位于旧城区的中心区域。当这片城区变得几乎不再适合作为步行道路穿行，人们开始思考重新设计、振兴这三处广场，成为更加宜人的场所。基于这种思想，主题很快就被确定下来，即"水之印迹"——水流经过的道路。专家们与普通民众在开放的氛围之中协同合作。戴水道设计公司为此制定了水景设计，雨水主要来自于教堂和市政厅之间广场上的雨水收集。那里拥有四级台阶，就像一大张折叠的毯子。如今，各个年龄段的人们都可以去追寻水流经过的印迹。在这个过程之中，人们也留下了自己的足迹，只要轻轻地踏入池塘之中，水流的方式便会随之改变。人们还可以使用置于台阶附近的造波设备改变水流的方式。高约5米的V形玻璃材质照明设备，如帽子般坐落于钢柱脚架之上，在黄昏来临之时将场地照亮。

1-3 Plaza with an interactive water carpet
广场上与使用者互动的水毯

4 Light columns reflect to the mayors house
光柱反射至市长办公室
5 People can play with waves and light
人们可以欣赏波光粼粼与灯影变幻
6,7 Aesthetic of water in flow patterns
水流动时的美感
8 Kids use the water as a stage of play
儿童将此处作为玩水平台嬉戏打闹

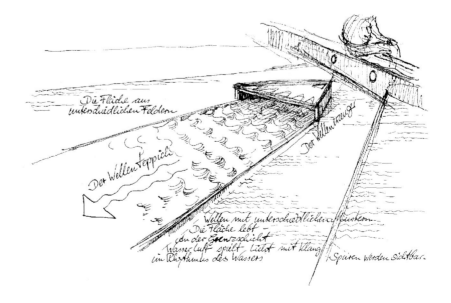

Heiner-Metzger-Plaza, Neu-Ulm, Germany
德国新乌尔姆，海纳-梅茨格广场

Location Neu-Ulm
Client City of Neu-Ulm
Area 2,600m²
Completion Time 2005
项目地址 新乌尔姆市
项目委托 新乌尔姆市政府
项目面积 2600平方米
建成时间 2005年

As the former main square fronting the train station, this plaza is a crucial building block in the inner city development scheme towards the Danube River and grounds of the federal garden show of 2008. Initial ideas for this project where developed by students as school projects. Atelier Dreiseitl took these ideas and developed them in an urban design vocabulary. Aesthetic and contemporary design was defined as the guiding design principles.

A one day workshop with the students at Atelier Dreiseitl gave opportunity to discuss the preliminary design. The students were able to directly influence the design, which brings with it a high acceptance of the final design and a sense of ownership for the students. The fundamental design idea is based on the differentiation of three distinctive zones: The field of diagonal pavers is interrupted by a field of sand and a permeable surfacing. From this, the tree canopy emerges. A climbing wall, a table football and a large-scale chess board offer opportunity for play. A water feature, that was tested and developed in full scale models, is the eye-catcher of the plaza and is visible from a distance. Four meter high veils of water descend into a tranquil pool with a water depth of 5cm. As vertical elements, these veils serve as accents and buffer the intrusive streetscape from the plaza. The sounds of the water contribute significantly to the atmosphere of this urban plaza. Potable water is used for the water feature and is collected in a cistern. It is circulated and technically cleaned.

作为曾经的火车站前主广场，该场地成为面向多瑙河开发和2008德国联邦花园展内城发展规划中一个关键的建筑街区。项目的最初想法源于学生的课程作业设计。戴水道设计工作室采纳了这些想法，并将之以城市设计的语汇发展深化。艺术性、现代化被定义为此项目的设计准则。

戴水道设计公司举办的为期一天的学生研讨会为初步设计讨论提供了机会。通过这样的交流过程，学生的设计想法能够直接影响设计，这便带来了一个具有高度认可程度的最终设计，对于学生来讲也具有强烈的归属感。基本的设计思想来源于三个特定区域的差异：斜纹铺装的场地被沙地、以及透水性铺面中断。由此，几株树木被种植于此。攀爬墙、桌式足球和大尺度的国际象棋盘提供了游乐的机会。经过全比例模型测试、开发的水景，成为了广场的焦点，从远处就清晰可见。高4米的水帘降落进入一个水深5厘米的宁静水池中。作为垂直景观元素，这些水帘可被视为一种强调，同时能够在一定程度上为广场阻挡干扰性的街景。水声大大有助于提升城市广场的气氛。水景用水来源为饮用水，它们被收集于一个蓄水池中进行循环和技术清洁。

1. City plaza for Neu-Ulm with flexible use
   具有多种功用的新乌尔姆城市广场
2. Large-scale chess board offer opportunity for play
   大尺度的国际象棋盘区域提供了娱乐场地

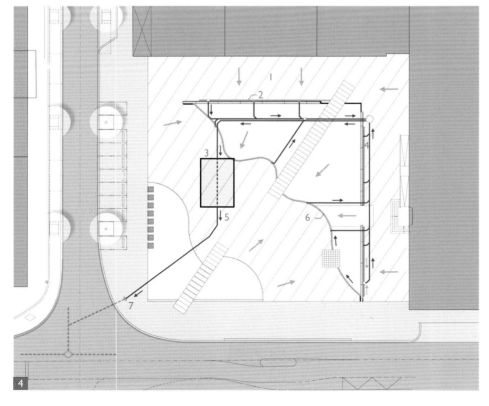

1. Flow direction
2. Cobble stone channel.
3. Subgrade infiltration basin
4. Collection channel
5. Overflow
6. Collection channel
7. Storm sewer connection

1. 流动方向
2. 卵石渠
3. 路基渗透基底
4. 收集渠
5. 溢流
6. 收集渠
7. 连接雨水排放管道

3 Plaza design with the water feature
水景广场设计
4 Stormwater drainage and infiltration
雨洪排水和渗透示意图
5 Neighborhood event are hosted often
社区活动经常在此举行
6-8 Activities for different generations
适合不同年龄段人群的活动

Water Management in McLaren Technology Center, London, UK
英国伦敦麦克拉伦技术中心水管理项目

Location London, UK
Client McLaren
Architect Foster&Partners
Partner Amac, Kier, WSP, Stahl
Area 1.62ha
Completion Time 2004
项目地址 英国 伦敦
项目委托 麦克拉伦
建筑设计 Foster&Partners
合作机构 Amac, Kier, WSP, Stahl
项目面积 1.62公顷
建成时间 2004年

1. Water Cascade, visible feature from inside
   从室内清晰可见的水帘
2. Cleansing biotopes are natural plant systems used to keep the water clean.
   生态净化群落为自然式植物系统，用以保持水体洁净。
3. An innovative water system cools the technical testing equipment in the building and provides an outstanding setting for the architecture.
   创新性的水系统为建筑内的技术检测设备冷却降温，同时也为建筑提供了优美的外观

As the United Kingdom becomes increasing divided through lines of drought and deluge, it was an essential challenge to site the new Formula 1 research and development centre into a protected natural site in a way that not just limited damage to the environment but actually positively contributes to it. Atelier Dreiseitl collaborated with Foster + Partner and the project team to create a smart, ecological water system which combines rainwater management and ecological restoration with the testing wind tunnels cooling system.

Stormwater run-off from the roof and parking lots is collected and stored in the lake. Lake water is circulated under the VIP access road through a natural planted biotope system, and from there to the building heat-exchanger and back out through a stunning 200m long cascade. Thanks to this natural cooling system the need for massive cooling towers was eliminated and thus the landscape character of the site protected. The adjacent stream is fed with clean water from the lake, a valuable ecological contribution in a country where streams are catastrophically running dry. The yin-yang of the building and lake blurs the line between interior and exterior spaces, making views to and from the building unforgettable.

由于英国的干旱和洪涝现象变得愈发极端和明显，因此通过一种不仅仅控制环境破坏、而且对环境有利的方式，将新的F1赛车调试研发中心设立在自然保护区内，就成为了一个极其重大的挑战。戴水道设计公司与福斯特建筑事务所以及整个项目团队合作创建了一个灵巧、生态的水系统，将雨水管理、生态保护与建筑冷却系统结合起来。

屋顶和停车场地上的雨水被收集起来储存于湖中。湖水通过自然生态群落系统在VIP通道下层循环，流至建筑热交换器，并从一个200米长、极具魅力的水瀑倾泻而下。此自然冷却系统的应用，避免了大型机械冷却塔的建造带来的环境破坏，保护了场地内的整体生态环境以及景观品质。清澈的湖水为临界河流提供了补给，缓解了溪流的季节性干涸，为地区的生态环境做出颇有价值的贡献。建筑物和湖体之间的阴阳造型概念模糊了室内外空间的界限，不论是望向建筑内部或是从室内远眺，其视觉感观都让人难以忘怀。

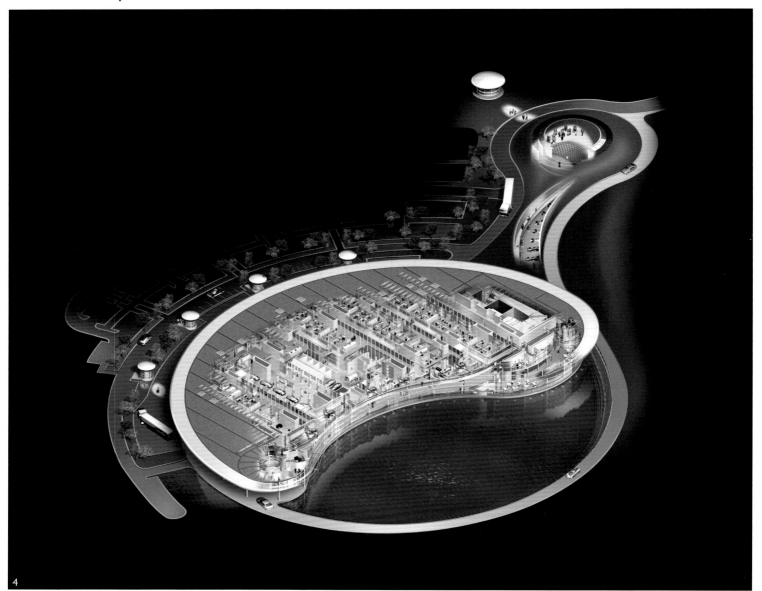

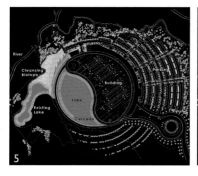 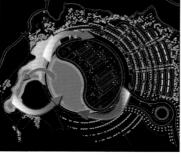 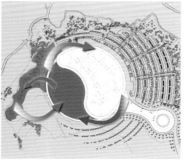 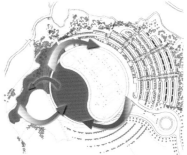

4 Building and formal lake as one design element
将建筑物以及形态规则的湖体作为一个设计元素
5 Water design and circulations cools the building
水体设计和循环对建筑进行冷却

6　Water cascade
　　水帘景观
7　The lake as an exciting entrance feature
　　湖体成为一个令人兴奋的出入口
8　Connection of water system and building
　　人工湖水面与建筑的交接处理

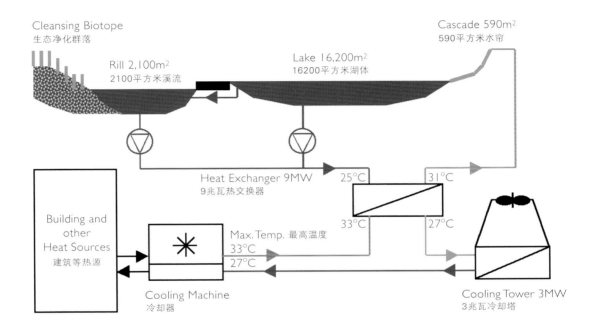

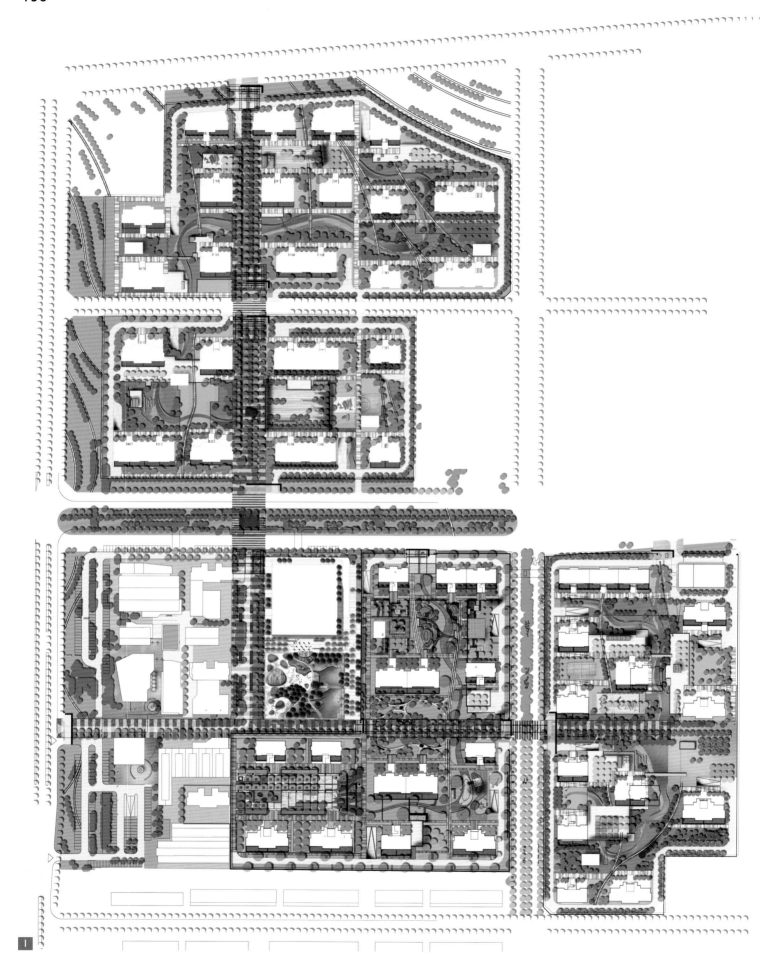

Mixed-Lifestyle Development, Changchun Vanke, China
中国长春万科综合生活区开发项目

Location Changchun, China
Client Changchun Vanke
Partner RSAA, Rheinschiene, WLA
Area 39ha
Design Time 2010-2011
Completion Time 2011
项目地址 中国 长春
项目委托 长春万科
合作机构 RSAA, Rheinschiene, WLA
项目面积 39公顷
设计时间 2010年-2011年
建成时间 2011年

Embedded in a picturesque landscape, shaped by wind and water, lies the 39-hectare project of Vanke Changchun in northern China. At the beginning of last century the plot was formed by the building of a huge diesel factory. In the 1920s several industrial buildings with brick facades were built, which were still in operation until 1990. As now a new mixed use was developed on that site, one of the most important questions was how this historic building structure could get integrated. First of all, analyzes of the building façade elements, foundations, towers, railway equipment and materials were undertaken to determine worth-while elements being preserved. After that a new plan was based on the existing geometric patterns that could match the requirements of use, density and character. Both concepts have been superimposed and a new district was developed, which is based on the existing industrial character of the relics. The open space concept with the large central park assimilates the elements of water, erosion and natural vegetation, and overlaid them with historic structures such as the old railroad tracks, but interpreted in a new design. Like a river, the park meanders through the strict grid and connects again scattered spaces and green oases. Meanwhile the first part of the residential estate is built (phase 1: 5.9ha.), e.g. the facade of the exhibition center with historic materials and techniques wall was rebuilt. The new high-rises with the historic structures are now a best seller and the further realization of the project is ongoing.

置于风景如画的北国，39公顷的长春万科项目坐落于风调雨顺的中国北方春城——长春。20世纪之初，该场地基址之上为一处巨大的柴油机厂房。20世纪20年代，大量的砖面结构工业建筑被建造，且一直使用至20世纪90年代。如今，场地之上进行了新的综合开发，所面临的一个重要问题即为如何对这些历史性建构物进行合理、有效的整合。首先，对于建筑立面元素、地基、塔楼、铁路设备和材料进行分析，以确定值得保留的元素。之后，基于符合场地使用要求、建筑密度和场地性质的现有几何地形图案，一个新的计划被制定。两个概念叠加在一起，根据遗迹的现存工业特征，对于新区域进行设计开发。拥有大尺度中央公园的开放空间概念吸收了水、冲刷和自然植被的元素，并将它们与历史性元素如旧铁轨结合在一起，以一种新的设计语言进行阐释。如河流般，公园蜿蜒穿过笔直的网格街区，并再次连接分散的城市空间和绿洲。同时，第一部分的住宅房产项目已经建造完成（一期项目：5.9公顷），其中，以具有年代感的材料和工艺墙壁铺设的会展中心墙体立面被重建。拥有历史性元素特征结构的新大楼如今已经成为了一个最好的卖点，而项目的进一步实践还在进行之中。

1 Overall plan of the mixed-use development
   复合型开发整体规划
2 Site old picture:
   an out-of-service diesel factory
   场地原有形态：废旧的柴油机厂房

3   Club plaza with reference to old railyard
    会所广场参照旧铁路站场而建
4   Gardens for residents
    住宅区花园
5   Pond with structured brickwall
    以砖墙堆砌的池塘
6   Enriched with lighting
    灯光装饰效果
7   Entrance plaza to the commercial area
    通往商业区域的入口广场

Mittelstrasse, Gevelsberg, Germany
德国格勒斯伯格市中央街设计

Location Gevelsberg
Client City of Gevelsberg
Partner MWH Engineers
Area 1.6ha
Design Time 2004-2007
Construction Time 2006-2008
Completion Time 2008

项目地址 格勒斯伯格市
项目委托 格勒斯伯格市政府
合作机构 MWH Engineers
项目面积 1.6公顷
设计时间 2004年–2007年
施工时间 2006年–2008年
建成时间 2008年

Mittelstrasse is Gevelsberg's main street, yet before the competition was a mess of traffic. The renovated street design is simple, modern, and pedestrian oriented while allowing for a smart vehicle circulation concept. The historic facades of the buildings define the space and the character of Mittelstrasse. Street trees are placed carefully to highlight this architecture and preserve the dynamics of the space. The pedestrian way is defined by striped paving which gives structure in a simple clear pattern. It is a tranquil atmosphere that is flexible for the many uses of a public street. At important intersections the street becomes an open plaza where cars and pedestrians share the space equally. This slows traffic and creates even more opportunity for public activities. Rainwater is collected from adjacent roofs and stored for use in 3 playful water features in pocket-plazas. The water experiences, landscaping, seating, architecture, businesses, and cafes make Mittelstrasse an urban living room for the residents of Gevelsberg.

中央街是格勒斯伯格市的主要街道，但在设计竞赛之前却交通混乱。整修街道的设计简单、现代、以步行交通为导向，同时引入智能化车行循环概念。富有历史韵味的建筑外墙定义了中央街的空间和格调。行道树被精致地排布，以突出建筑本身并保存该空间的动态特征。步行道路被定义为条纹铺路，赋予道路结构以简单、清晰的模式。这里气氛宁静，作为一条公共街道，可以灵活地满足多种用途的需求。在重要的路口区域，该街道则成为了一个开放广场，汽车和行人共享空间。这样的空间设置将交通流速降低，为公共活动创造了更多的机会。从相邻屋顶之上收集到的雨水被存储，并作为小型广场上的三个嬉水游乐设施的用水来源。水体验、园林景观、休息设施、建筑欣赏、商业活动以及分布众多的咖啡馆，将中央街打造成为了格勒斯伯格市居民们的一处城市客厅。

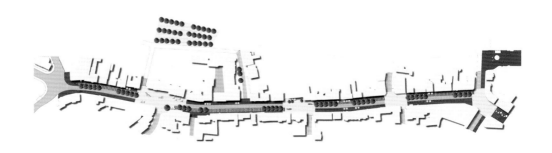

1　Urban design of the new main street
　　城市新的主要街道设计

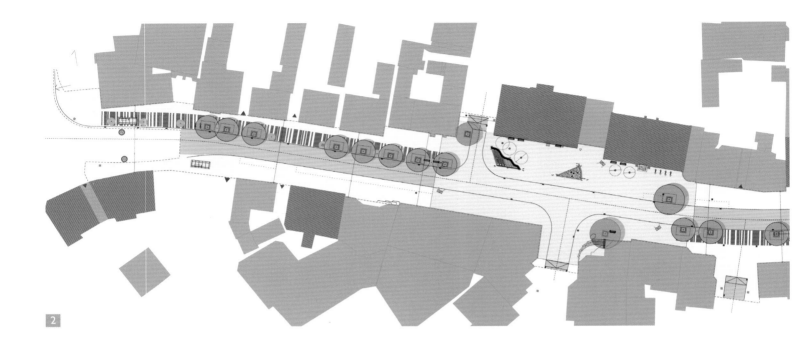

2

3

2 Detail plan for the streetscape design
街景细部设计方案
3 Plaza with interactive water feature
广场上互动式水景元素
4 Functional areas are smoothly seperated with pavement
功能区域平缓的与道路铺装分隔开来
5 Human scale city encourage for communication
人性化尺度的城市有益于交流活动
6 Unique pavement refers to barcode
形如条形码般独特的道路铺装

Integrated Stormwater Management Strategy for Ecological City

by Wu Che, Yang Zhao, Pan Yan

Introduction

Extensive mode of urban development causes severe environment interference and destroys ecological balance which are extremely difficult to be mitigated and recovered in the future. Urbanization in China has entered a critical stage, however there are even deeper and broader space for improvement. It has become an urgent, long-term, and arduous task to coordinate relationship between urban development and ecological environment.

People's concerns and requirements for safe and sustainable living conditions are rising rapidly with the improvement of living standard and civilization level. As a result, the rapid development pattern in the past which depends on high consumption of environmental resources cannot be continuing. The ecological civilization is becoming important issue to be concerned in urban development. In the 18th meeting of Chinese Central Committee (CPC), ecological civilization has been announced on the prominent position in the key national development program and the construction of ecological, natural "Beautiful China" has become an important national strategic objective. However, we should be clearly aware of that pollution, environmental damage, and even urban safety crisis are never stopped and are definitely needed to be urgently resolved.

Frequent water crisis happened in quite recent years raised challenge for traditional urban development mode. Severe water logging events in big cities like Beijing result in loss of large amounts of property, or even citizen's lives. Urban sewage treatment in Beijing City and other Chinese cities has covered 90% percentage of city area, however the stormwater runoff is still seriously polluting river and lake water system, leading to widespread eutrophication and big amount fish death. As for urban water shortage, rainwater resource in flooding season is in a huge waste with groundwater level's continuing decline, leading to a more fragile ecological environment. These contradictions are prominent and related one another. It has become an inevitable requirement to build up a sustainable drainage system for realizing eco-city vision.

Stormwater Management Strategy for Ecological City

Recognizing limitations of traditional drainage system, and for meeting the requirements of safe, comfortable and ecological environment in urban areas, western countries who have ever experienced the same confusions finally set up a series of technical systems. The most representatives are such as Best Management Practices (BMP), Low Impact

Development (LID) in United States, Sustainable Drainage Systems (SUDS) in the UK, and Water Sensitive Urban Design (WSUD) in Australia. Although these systems are slightly different in definition and interpretation, all of them are seeking for a natural, sustainable and multifunctional pattern instead of traditional development model. The aims of these systems are trying to solve urban stormwater problems, and to realize well-functioning hydrological cycle and protect livable city environment.

During decade's continuous exploration, research, and practices in stormwater system construction, we developed multi-scale, multi-disciplinary, multi-objective planning and design concepts (Figure 1) which are tested in various projects. Regarding to considerations such as difficulties in dealing with urban water, eco-city development requirements, social and natural conditions, existing urban planning characteristics and policy approaches, we do believe stormwater management strategy in Chinese urban area should meet the following key points:

•   Integrated stormwater management strategy should keep the restoration of hydrological cycling as the highest goal and meanwhile with sub-goals which cover contents in realizing urban hydrological safety, water pollution prevention, water resource utilization and designing ecological waterscape. This strategy mainly emphasizes sustainable development of the urban hydrological system.

•   To protect and utilize site natural resources, such as wetlands, ponds, natural ditches, and try to avoid negative interferences on site landscape and ecological elements.

•   By combining new stormwater facilities such as LID (Low Impact Development), GSI (Green Stormwater Infrastructure) with traditional infrastructure and utilizing ground and underground space, to develop comprehensive stormwater control system from the starting, procedural and ending points.

•   It needs to emphasize the importance of special planning in rainwater controlling and utilizing. Combining new ideas, goals with technical measures, special planning integrates relationships between keeping water quality, preventing water logging and optimizing water use. At the same time it is useful to connect landscape design, space utilization, and traffic system planning.

•   It is necessary to coordinate the demand for green building and eco-city construction. By designing and optimizing water system in green building, drainage systems and water environment in eco-city are well-integrated.

•   To provide support for technical system by developing related policies and regulations, establishing multi-disciplinary coordination mechanisms, improving management and maintenance techniques.

Waterscapes Innovation | 146

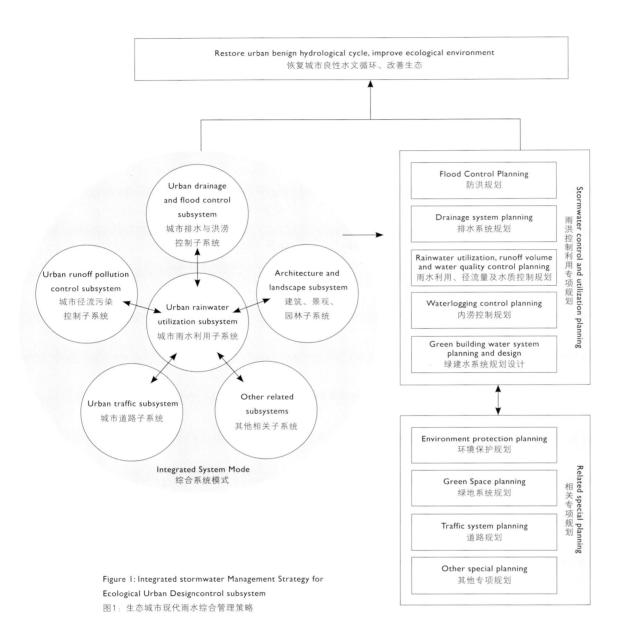

Figure 1: Integrated stormwater Management Strategy for Ecological Urban Designcontrol subsystem
图1：生态城市现代雨水综合管理策略

In practice, according to different characteristics and requirement of each project, specific treatment of modern stormwater management system is various, however, the common object is to improve ecological environment and restore well-running urban hydrological cycle. There are some successful projects such as Vanke Shenzhen Dameisha headquarter as a model of green building (LEED Platinum certification of the United States and three-star LOGO of Chinese Green Building certification). It is featured by the coordination of landscape design, green building assessment and stormwater management system. Special ecological stormwater system is built in process of site design for carrying, infiltrating, purifying, and reusing rainwater resources. Water system planning for Giant Panda Research Center reconstruction project in Sichuan Wolong / Dujiangyan takes into consideration conservation and utilization of key ecological elements (original gullies, wetlands, etc), geological disasters solutions, rainwater reuse and emission, and water system construction for animal and people. Another example is a large-scale residential project -- Beijing Oriental Sun City. It is entirely designed to drain, reuse and do flooding control through ecological stormwater system. In nearly ten years, it never appears to be water logging after heavy stormwater in Beijing and has been successfully in creating a rich ecological waterscape with maintenance costs saving. In other projects as Beijing Changping Science and Technology City, Dalian ecological Technology Innovation City, land use, landscape design and water system are integrated by interdisciplinary collaboration in Planning, Landscape design, Water supply and drainage. Thus, urban planning and water system construction are closely connected in aspects of planning, design, management and ecological protection.

Conclusion

Integrated Stormwater Management Strategy not only meets multi-objectives like flooding control, rainwater resources regulating, peak runoff storing and adjusting, water purification, water environment protection, but also has further influences on areas such as purifying air condition, energy saving and emission cutting, alleviating urban heat island effect, enhancing carbon sequestration, improving land value and saving investment input. It is also beneficial to provide the public natural landscape and livable environment with aesthetic and ecological features.

Therefore, it is significant and urgent to build well-functional hydrological cycle into practice. It is a kind of basic requirement of sustainable urban development and ecological construction. It should be clearly recognized and become a consensus by professionals in urban planning, architecture, landscape, municipal affairs, water conservancy, environmental protection, etc. It needs to be practiced and promoted in projects of green building construction, refurbishing, urban district transformation and ecological city construction. The concept of creating well-functional hydrological cycle would be regarded as basic guidance for infrastructure construction and urban development towards sustainability, which providing strong support in the process of realizing Beautiful China.

生态城市雨水综合管理策略
车伍、赵杨、闫攀

一、引言

粗放型的城市发展模式对生态环境造成的剧烈干扰甚至破坏，缓解与恢复都极为困难。中国城市化已进入关键发展阶段，未来仍具有大幅提升空间，协调城市发展与生态环境的关系是一项紧迫、长期而艰巨的任务。

随着生活水平和文明程度的提高，人们对环境安全及生态的关注与要求也在迅速提升。正因为此，过去那种以牺牲环境和生态为代价的高速发展已难以为继，生态文明成为城市发展必须兼顾的重要方面。党中央在十八大特别将生态文明放在"五位一体"的突出地位，建设生态、自然的美丽中国已经成为国家重要战略目标。但在积极建设生态城市的同时，我们也应客观、清晰地意识到，城市开发造成的污染、环境破坏、甚至城市安全危机仍不断发生，亟待解决。

近年来，国内城市频发的复杂水环境危机已经对传统发展模式提出了疑问。许多城市不断爆发的严重内涝事件，造成巨额财产损失，甚至付出生命的代价；北京等发达城市污水处理率超过90%，但雨水径流依然严重污染河湖水系，导致普遍的富营养化和大量鱼类死亡；城市缺水的同时，汛期雨水资源却大量流失，地下水位持续下降，生态环境更加脆弱。这些矛盾突出且互相关联，构建可持续的雨水系统已成为实现生态城市愿景的必然要求。

二、生态城市雨水综合管理策略

面对传统排水系统的局限，以及城市发展对安全、环境舒适性以及生态的要求，发达国家在经历同样的困惑时通过不断探索，逐渐发展出一系列雨水控制利用技术体系，代表如美国的最佳管理措施（Best Management Practices，BMP）、低影响开发（Low Impact Development, LID），英国可持续排水系统（Sustainable Drainage Systems SuDS），澳大利亚水敏感性城市设计（WSUD）等。这些体系虽在名称和内涵上有所差异，但均力求改变传统开发模式，以可持续的、与自然充分和谐的及多功能的手段解决城市雨洪问题，并以恢复与构建城市良性水文循环、保护生态环境为最高目标。

在我们十余年不断探索、研究、践行中国雨水系统建设的过程中，多尺度、多专业、多目标的综合规划设计思路（图1）始终贯穿每项研究课题与工程实践任务，并通过不同类型的实际项目进行实践。总结多年经验，结合中国城市发展中面临的突出水问题、生态城市建设要求、社会与自然条件、规划体系、政策等情况，适合于国情的雨水综合管理策略的基本特征应有：

* 雨水综合管理策略应将恢复水文循环作为生态城市雨水系统的最高目标，子目标涵盖或衔接城市安全、污染防治、资源利用、景观生态等多方面内容，强调水系统的可持续发展。
* 保护与利用场地自然资源，如湿地、坑塘、自然沟渠等，尽量避免开发对场地地貌及生态元素的干扰。
* 结合LID、GSI等新型雨水设施及传统基础设施，利用地上与地下空间，从源头、中途、末端综合构建控制系统。
* 强调雨水控制利用专项规划的重要性。伴随新的理念与目标，雨水技术措施的落实，需要运用专项规划整合雨水相关水质、内涝、利用等各个子系统的关系，同时衔接景观、空间利用、道路系统等相关规划。
* 协调绿色建筑与生态城市的建设需求，通过绿色建筑水系统的设计及优化，综合构建生态城市雨水系统与水环境。
* 制定政策与法规、建立多专业协调机制、管理维护等非工程措施为技术体系提供支持。

在实践过程中，根据不同项目特点和需求，现代雨水管理系统的实施方案也有很大区别，但主要目标都是改善生态、恢复城市良性水文循环。一些较为成功的案例，比如万科大梅沙总部作为绿色建筑的典范（获得美国LEED铂金认证以及中国绿色建筑三星标识），通过协调景观设计、绿色建筑评定及雨水系统建设需求，在场地设计阶段进行了生态雨水系统专项规划，利用生态雨水措施疏导、下渗、净化、回用雨水资源；卧龙／都江堰大熊猫研究中心灾后重建项目水系统规划设计协调了场地重要生态要素的保护与利用（原生态冲沟、湿地等）、地质灾害的处理、雨水回用及排放、动物与人类水系统建设等多种复杂目标；北京东方太阳城大型住宅小区完全通过生态雨水系统组织排放、回用、洪涝控制，在不使用雨水管道的情况下，近十年来历经暴雨未出现水涝，并成功营造丰富的水环境生态景观，节省投资；北京昌平科技城、大连生态科技创新城，通过规划、景观、给水排水等跨学科合作，统筹土地利用、景观设计与水系统的关系，在规划、设计、管理、生态保护等各方面建立城市规划与水系统建设的紧密联系。

三、结语

现代雨水综合管理策略不仅可实现洪涝控制、调蓄利用雨水资源、滞留调节径流峰值、净化水质、保护水环境等多目标，更兼具净化空气、节能减排、缓解城市热岛效应、增强固碳作用、土地增值、节约投资，为市民提供具有美学和生态功能的自然景观和宜居环境等许多功能。

通过综合管理雨水资源构建城市良性水文循环，刻不容缓且意义重大。它是城市可持续发展和生态建设要求下的一种基本理性，是城市基础设施多功能、生态化、集约型发展趋势的典型体现，应被城市规划、建筑、景观、市政、水利、环保等不同行业的专业人员认知并达成共识，并通过绿色建筑、既有建筑和城区改造、生态城市等不同载体来实践和广泛推广，由点及面地引导城市发展和基础设施建设走向可持续，为建设美丽中国提供有力的支持。

E01 | Urban Park 城市公园

Tanner Springs Park, Portland, USA
美国波特兰坦纳斯普林斯公园

Location Portland, USA
Client City of Portland
Partner GreenWorks PC
Area 4,000m²
Completion Time 2005
Awards ULI Open Space Award Finalist 2011; ASLA Oregon Chapter, Merit Award Landscape Design 2006
项目地址 美国 波特兰市
项目委托 波特兰市政府
合作机构 GreenWorks PC
项目面积 4000平方米
建成时间 2005年
所获奖项 2011年城市土地学会开放空间设计奖入围、2006年美国景观设计师协会俄勒冈州景观设计优胜奖

Formerly a wetland, the Pearl District was bisected by Tanner Creek and sided by the broad Willamette River. Rail yards and industry first claimed and drained the land. Over the past 30 years, a new neighborhood has progressively established itself – young, mixed, urban and dynamic, today the Pearl District is home to families and businesses. With surgical artistry, the urban skin of one downtown block, 60m x 60 m (200 ft by 200 ft) is peeled back to create a new city park. Stormwater runoff from the park block is fed into a natural water feature with a spring and natural cleansing system. The 'Art Wall' recycles historic rail tracks, oscillating in and out and inlaid with fused glass pieces hand-painted with nature images by Herbert Dreiseitl. Ospreys dive into the water, art performances unfold on the floating deck, children splash and explore, and others take quiet contemplation in this natural refuge in the heart of the city. An intense community participation and a stakeholder steering group means that this park is the realization of the dreams and hopes of local people.

波特兰珍珠区基址原来为一片清泉滋润的湿地，被坦纳河(Tanner Creek)从中划分开来，与宽广的威拉麦狄河（Willamette River）相邻。铁路站和工业区首先占用了这片土地，并伴有场地排水要求。在过去的30年里，一个新的社区被逐步建成，它象征着年轻、综合、大都市和活力。今天的珍珠区已经成为了商业和居住区域。在一个市区繁华地带大约60米×60米的地方，重新塑造一个崭新的城市公园。从公园街区收集的雨水汇入由喷泉和自然净化系统组成的天然水景。从铁路轨道回收的旧材料被重新利用并建造公园中的"艺术墙"，唤起人们对于历史铁路的记忆，而波浪形的外观设计则能够给人以强烈的冲击感。戴水道设计公司的创始人赫伯特·德赖塞特尔先生本人通过手绘，将这里曾经生存的生物图案绘制于热熔玻璃上，并镶嵌在"艺术墙"内。在这个繁华的市中心地带，生态系统得到了恢复，人们居然可以看到鱼鹰潜入水中捕鱼。在甲板舞台上可以尽情地表演各种文艺活动，孩子们来到这里玩耍、探索自然奥秘，而另外一些人们则可以在这片自然的优美秘境中充分享受大自然芬芳、进行无限的冥想。深入的社区参与和地产调查显示，这个公园是当地人们实现梦想和希望的地方。

1 As a key urban park for the new Pearl District
 新珍珠区主要的城市公园
2 Osprey diving into the water
 鱼鹰潜入水中

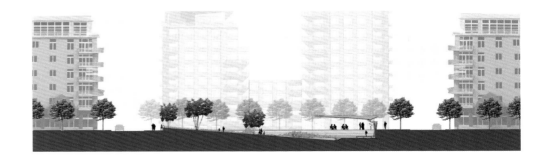

3 Even educational activities are held
在此场地举行教育活动

4 Cultural events find a good spot
文化活动寻找到了一处好的场地

5 Recreation and experience of urban nature
休闲和体验城市自然

6 Design concept of an urban wetland
城市湿地的设计概念

7 Stormwater is collected from roads
道路上的雨水收集

Plan:
1. Trees
   Street trees:
   Greenvase zelkova
   Interior trees:
   Oregon white oak
   Red alder
   Big leaf maple
2. Cobbled paths
3. Spring
4. Meadow
5. Benches
6. Slow track sidewalk
7. Fast track sidewalk
8. Rain pavilion
9. Art wall
10. Boardwalk
11. "Floating" pontoon
12. Wetland
13. Water
14. Seating steps

平面图：
1. 树木
   沿街树木：
   榉树
   公园内部树木：
   俄勒冈白橡树、
   赤桦木、
   大叶枫
2. 鹅卵石小径
3. 泉水
4. 草地
5. 长椅
6. 慢行边道
7. 快行边道
8. 雨亭
9. 艺术墙
10. 木栈道
11. 浮桥
12. 湿地
13. 水体
14. 休息阶梯

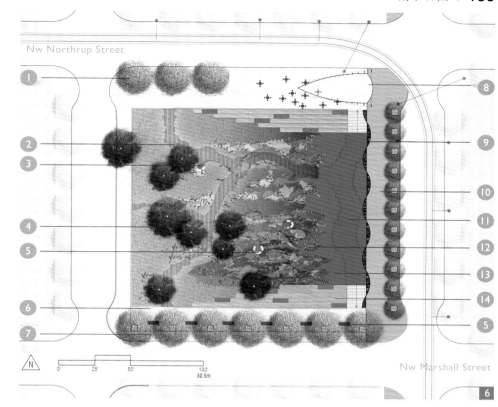

Plan:
1. Rainwater pavilion
2. Collecting from pedestrian zone
3. Overflow
4. Collecting from streets
5. Stormwater cleansing

平面图：
1. 雨水亭
2. 步道区域收集雨水
3. 溢流
4. 街道收集雨水
5. 雨水净化

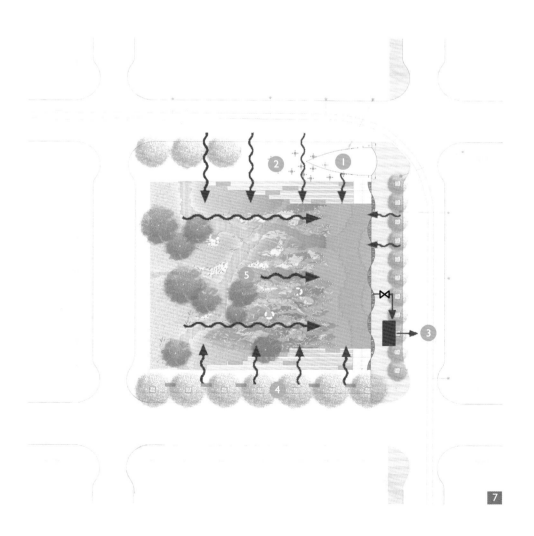

Queens Botanical Garden, New York, USA
美国纽约皇后植物园

Location New York, USA
Client New York Queens Botanical Garden
Partner BKSK Architects,
CDF Landscape Architects
Area 16ha
Construction 2005
Awards American Institute of Architects/
COTE Top Ten Green Project 2008
项目地址 美国 纽约
项目委托 纽约皇后植物园
合作机构 BKSK Architects,
CDF Landscape Architects
项目面积 16公顷
建成时间 2005年
所获奖项 2008年美国建筑师协会环境委员会
绿色环保项目奖前十名

Just over the river from Manhattan, in the ethnically most diverse borough of the United States, lies the Queens Botanical Gardens. Atelier Dreiseitl teamed with Chicago-based plant experts Conservation Design Forum and local architects BKSK to develop a people inspired master plan.

The first phase included the construction of the new administration building and entrance including a building integrated rainwater recycling system visible from the entrance. Participation of the ethnically diverse local community and Garden volunteers and staff Design workshops drew out the invaluable contributions of the Garden staff and local residents.

The masterplan accommodates 0% run-off of stormwater for a 1 in 100 storm. Attractive stormwater detailing, such as 6,000-square-feet green roof on the new administration building and the reuse of storm water run-off for cultural water features and irrigation, give the garden a unique, water related character. Gray water is captured, cleansed and reused for low-contact irrigation. A new 46,000-square-feet "green parking garden" demonstrates how environmental design can be both effective and beautiful.

与美国纽约曼哈顿隔河相望、在美国种族最多样化的城区之中，坐落着纽约皇后植物园。德国戴水道设计公司联手芝加哥地区植物专家机构——保护设计论坛以及当地建筑事务所BKSK共同开发人性化的规划设计。

此设计的第一阶段包括建造新的行政大楼和入口，其中包括从入口处可见的建筑整合雨水循环系统。在当地多族裔社区、植物园志愿者和工作人员的参与下，设计团队充分采纳植物园员工和当地居民的宝贵建议并将其融入设计之中。

该设计可以容纳百年一遇暴雨的全部雨水径流。引人注目的雨水处理细节，例如新的行政大楼之上6000平方英尺的屋顶花园，以及将雨水径流进行重新利用、用于文化性水景和灌溉用水，为植物园带来独特的水质景观。中水被收获、清洁并循用于低接触性灌溉。一处新建46000平方英尺的"绿色停车花园"展示了环境设计如何具有实效、且美丽迷人。

1 Runoff as an exciting feature
  跌落水体成为了一处独特的景致
2 People exercise in the garden
  人们在公园中锻炼

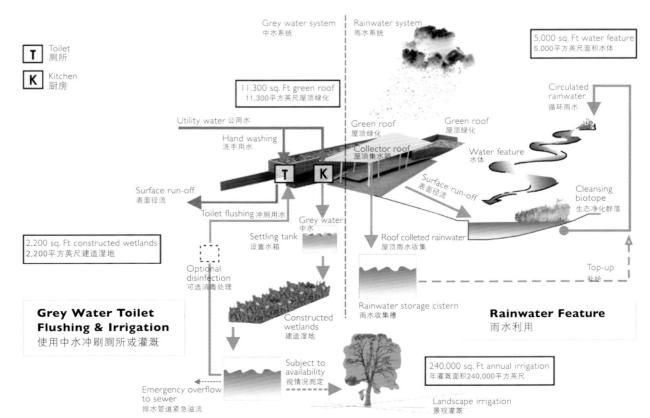

3 Water features and building as one concept
水景和建筑融为一个设计概念

4 Comprehensive water system use symbioses
综合的协同水系统应用

5 Watercourse connect old and new
河道将新旧场地相连接

6 Landscape integrates the building harmoniously
景观与建筑物和谐共存

7 Green roof as a sloped garden in the building
建筑上的绿色屋顶成为一处倾斜的屋顶花园

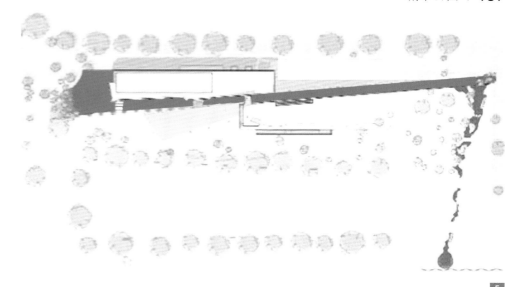

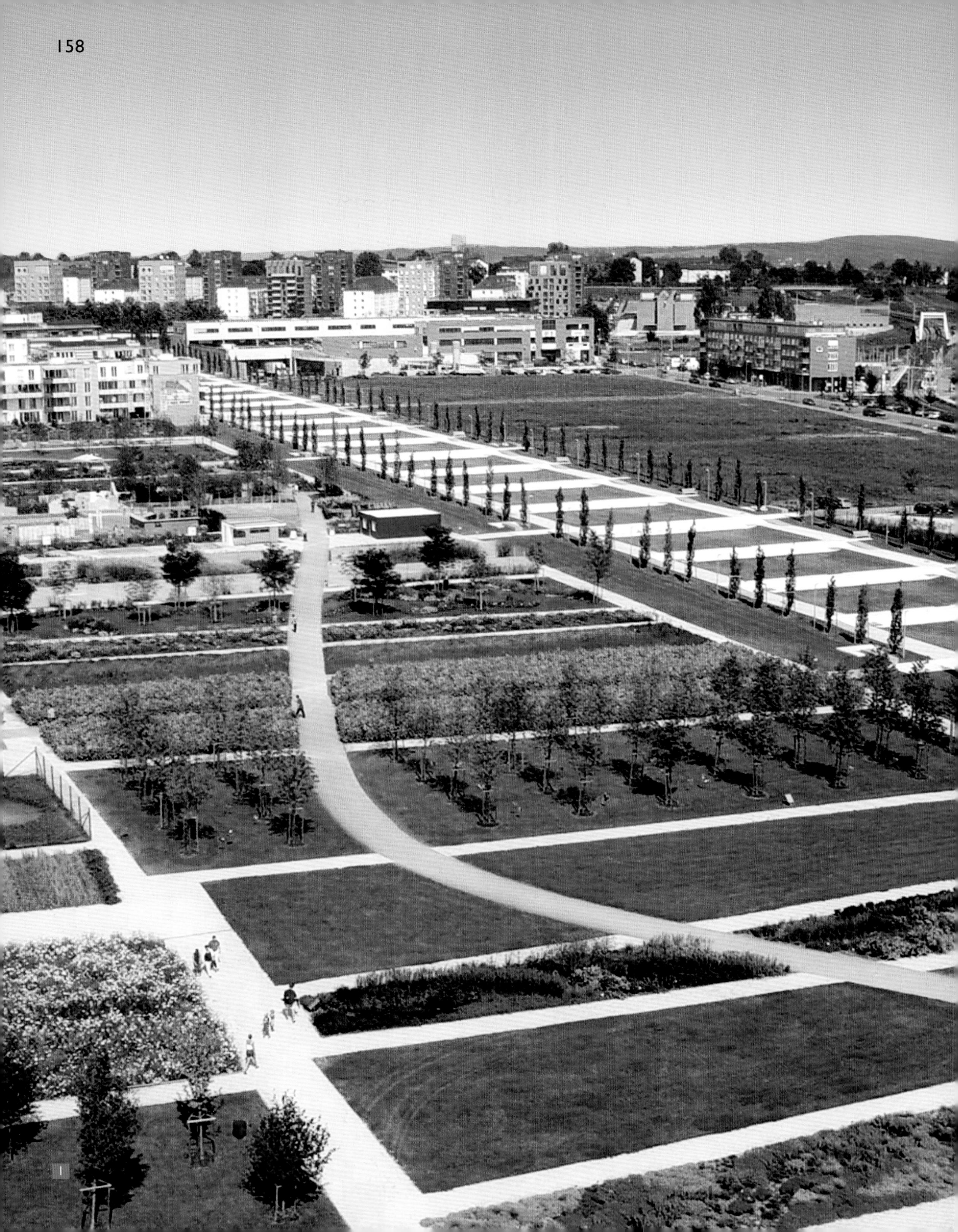

1

Scharnhauser Park, Ostfildern, Germany
德国奥斯特菲尔登沙恩豪瑟公园

Location Ostfildern, Germany
Client City of Ostfildern
Partner Janson&Wolfrum Architects
Area 150ha
Construction 1996-2004
项目地址 德国 奥斯特菲尔登
项目委托 奥斯特菲尔登市政府
合作机构 Janson&Wolfrum Architects
项目面积 150公顷
施工时间 1996年-2004年

Stuttgart is an expanding urban center in southern Germany. A disused army base offered the opportunity to create a new town along the local city rail link. The main challenge was how to keep rainwater run-off at the same rate after development as before, as the site lies close to a tributary feeder to Stuttgart's main river, the Neckar, and is on steep, heavy clay hill. Scharnhauser Park near Ostfildern is laid out over 150 hectares, and is the largest urban development scheme in the Stuttgart area in the early 21st century. The scheme was combined with the State Garden Show which has helped promote and phase in construction of the project. The urban layout is inspired by an innovative and cost-effective rainwater management system which keeps rainwater on the surface in attractive streetscape detailing, and retains and filters run-off in a series of stepped technical multi-purpose parks and tree boulevards. Not just the newts and bats are happy, this ecological infrastructure offers great places for children to play, adults to live and makes a valuable contribution to the greater watershed management of Stuttgart.

斯图加特是德国南部一座不断扩张的中心城市。一处废弃的军事基地为沿当地城市铁路线创建一座新城提供了时机。因为场地临近斯图加特主要河流——内卡河的支流入口，且场地地形陡峭、土质黏稠，其设计主要难点在于如何在开发建设后保持与之前相同的雨水径流速率。靠近奥斯特菲尔登的"沙恩豪瑟公园"占地150公顷，是21世纪初期斯图加特地区最大规模的城市开发计划。该计划与"国家园林展"（State Garden Show）相结合，以提升项目知名度、确定项目的分阶段实施。城市设计布局以创新性、成本收益合理的雨水管理系统为特征，该系统将雨水保留于优美的街景细节之中，通过一系列阶梯状、技术性、多功能公园和林荫道路保存并过滤雨水径流。在这里，不仅鸟兽虫鱼备感亲切，生态基础设施的建立也为孩子、成人提供了极佳的玩耍、生活空间，为斯图加特城市更加美好、有效的流域管理做出了有价值的贡献。

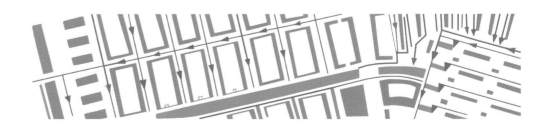

1 Development during the Garden exhibition
园林展期间的开发

E03 | Urban Park | 160

2 Urban parks for retention
   具备蓄流功能的城市公园
3 Detention areas
   with multiple use
   多功用的滞留区
4 Flexible with
   recreational purpose
   灵活使用的休闲功能
5 Shared space with
   integrated tunnels
   整合了排水管道的公共空间
6 Landscape staircase
   with drainage system
   具备排水系统的景观阶梯
7 Overall stormwater system
   整体雨水系统设计方案

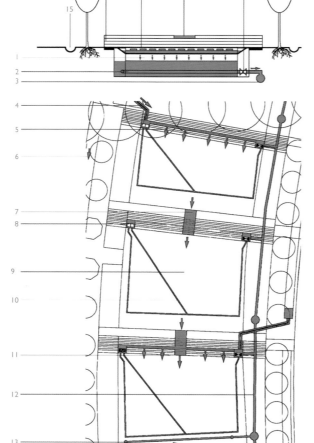

Legend:
1. Top soil
2. Trench drain and drainage
3. Rain water canal
4. Inlet
5. Narrow slot weir
6. Collecting trench for roof- and rainwater
7. Steps overflow
8. Reduced outlet manhole incorporating gravel trench detention
9. Infiltration swale
10. Drainage net
11. Reduced outflow
12. Rain water canal
13. Emergency overflow
14. Outlet to retention area east
15. Trench
16. Steps
17. Narrow slot weir
18. Steps overflow
19. Reduced outlet/orifice

图注：
1. 表层土
2. 沟渠排水
3. 雨水渠
4. 进口
5. 狭槽堰
6. 屋顶雨水收集沟渠
7. 阶梯溢流
8. 减量的排水井并入砾石滞留沟槽
9. 入渗洼地
10. 排水网
11. 减少外流
12. 雨水渠
13. 紧急溢流
14. 出水口引流至东侧滞留区
15. 沟渠
16. 阶梯
17. 狭槽堰
18. 阶梯溢流
19. 限流口/孔

城市公园 | 161

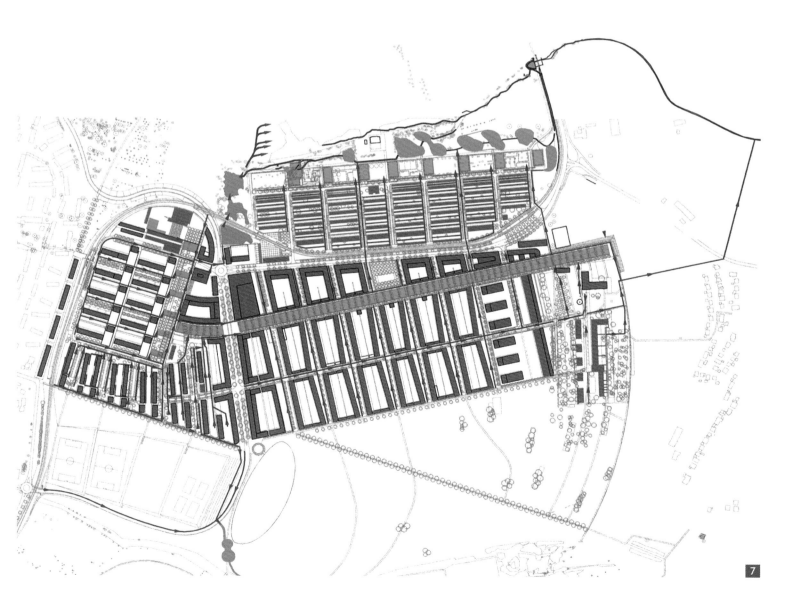

Green Roof, Chicago City Hall, USA
美国芝加哥市政厅屋顶花园

Location Chicago, USA
Client City of Chicago
Partner William McDonargh&Partners,
CDF Landscape Architects
Area 3,600m²
Construction 2001
项目地址 美国 芝加哥
项目委托 芝加哥市政府
合作机构 William McDonargh&Partners,
CDF Landscape Architects
项目面积 3600平方米
施工时间 2001年

Chicago is one of the five major American cities taking part in the environmental authority´s 'Urban Heat Island Initiative' pilot project. This is the United States' attempt to reduce temperatures and the smog level in several cities. Roof planting is one of the key elements of this programme, which is also intended to relieve the overloaded sewers when the rainfall is heavy.

There are very few roof planting experts in the USA because of lack of experience. For this reason Atelier Dreiseitl was invited to join an American planning team in 1999, and commissioned to produce a design for the city hall of Chicago. What has emerged on top of the city hall is a lightly contoured landscape, planted on shallow substrate with varieties of sedum and on a deeper one with trees and shrubs. It is possible to walk around the city hall roof on a curving path. Parts of the roof were removed for statical reasons before the planting, and provision of water for the roof plants was dealt with as part of this process. Rainwater from the penthouse which is built directly on the city hall, and higher, is stored in several small tanks and taken to the plants when needed. In Chicago the project is the first example of the fact that the planting of roofs is worthwhile, even from the point of view of economical and sustainable water management.

作为美国五个主要城市之一，芝加哥市参加了美国环保署的"城市热岛倡议"试点项目。该项目即在美国一些城市之中，试图降低城市气温和烟雾水平。屋顶绿化是此项目方案之中的关键要素之一，同时也有助于减轻暴雨之时下水系统的过度负荷。

因缺乏实践经验，美国屋顶种植方面的专家数量有限。为此，在1999年，德国戴水道设计公司受邀加入一个美国规划小组，并受委托为芝加哥市政厅设计一个屋顶花园。出现在市政厅顶部的是一个轻质结构景观，浅层基质上种植多品种景天属植物，而稍深基质层上则生长乔灌木。弯曲道路的设置使得人们能够行走在市政厅的屋顶之上。开始种植前，部分屋顶因静置的原因从设计中被移除，在此过程中，对屋顶植物的供水处理也被同时进行。从直接建造于市政厅屋顶之上的构筑物收集的雨水被存储于几个小水塔中，在需要时用于植物灌溉。该项目成为芝加哥市第一个实例，证明了从经济角度和可持续水管理的角度，屋顶绿化种植是可行并值得的。

1 City Hall by night with green roof
  拥有屋顶花园的市政厅夜景
2 Green is mitigating Extrem weather
  城市立体空间绿色植被
  有益于缓解极端气温
3 Even bees find their Way to make honey
  蜜蜂也找到了采蜜的地方
4 Roof top plan
  屋顶方案设计

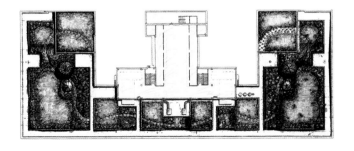

Maybach Center of Excellence, Stuttgart, Germany
德国斯图加特迈巴赫卓越中心

Location Sindelfingen, Stuttgart, Germany
Client Daimler Chrysler AG
Partner Kohlbecker Architects
Area 1,050m² (Central Catchment)
Construction 2002
项目地址 德国 斯图加特 辛德尔芬根
项目委托 戴姆勒克莱斯勒公司
合作机构 Kohlbecker Architects
项目面积 1050平方米
施工时间 2002年

1 Sound is made by a cascade
  跌水响声
2 Cleansing cascade visible through
  净化跌水清澈见底
3 Building stands in the water
  建筑物伫立于水体之中
4 Rich vegetated banks
  植被繁茂的水岸

DaimlerChrysler wanted to create an atmosphere of distinct elegance in the architecture and ambience of their new Center of Excellence near Stuttgart, reflecting that something extra which defines their new line of Maybach cars. Water is the medium which connects the new Center of Excellence to the existing Client Center. The water has a mirroring effect, the Center of Excellence appears to glide over the water.

To approach the main reception area, visitors cross a narrow bridge which brings them across into the inner area. The waterbody is fed by rainwater collected from the roofs. Without chemicals, in a system which is both ecologically and economically sound, water is circulated in two separate cycles, one for cleansing and the other to reduce stagnant zones in the water body. Wall elements and window openings give articulated views from the water into the building and vice versa.

The soft natural shoreline with attractive planting contrasts with the simple formality of the building, a pleasing juxtaposition which gives the Center a special character.

戴姆勒克莱斯勒公司希望在建筑形式和氛围上，将位于斯图加特附近的新"卓越中心"营造出一种独特的典雅气氛，以彰显其新型迈巴赫豪华轿车与众不同的品质。在此，水成为了连接新的"卓越中心"与现有客户中心的媒介。在水体镜像效果的映衬下，"卓越中心"仿佛滑行于水面之上。

穿过一条窄桥，游客便可以进入主接待区、直达建筑内部区域。湖体水源来自屋顶收集雨水。湖水中不含化学制剂，无生态污染、且经济实惠。水体在两套独立的系统之内循环流通，其一用于清洁，另外一个则用于减少湖中的淤积区域。无论从水面望向建筑、或是反向，墙体元素与开窗设置都为游客提供了清晰的视野。

柔美的自然式水岸线与形式简洁、规则的建筑形成对比，令人愉悦的整体环境赋予"卓越中心"别样的特色。

E05 | Urban Park | 166

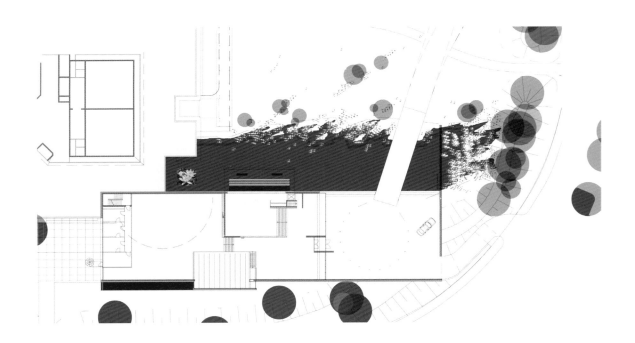

5 Stone blocks create under water landscape
石块创建了水下景观

6 Overall plan
整体规划方案

7 Patio for events and the sloped bridge
活动场地和坡行桥

Think Global Act Local
by Prof. Wolfgang F. Geiger

Water is neither inexhaustible nor invulnerable. But the intensity with which it is used today tends to ignore these facts, as we increasingly exploit and pollute this gift of nature that is so essential for life. If we do not want to have to dig for our own water in future, we must think cooperatively, decentralize, and establish autonomous systems for water use at a local level.

There is no other natural resource on which mankind makes such heavy and complex demands as it does on water. Although it is not renewable in part, we neglect it far more than other resources – just remember how oil exploitation was coordinated internationally. In contrast with this, we treat water as though it were inexhaustible. Philosophy, science and technology have contributed to this mistaken assessment.

On the whole, people prefer and have always preferred to establish towns near water. It can then be exploited directly, it is a transport medium that promotes trade, and it contributes to the well-being of the inhabitants. Water in a town fulfils cultural, architectural and social functions. The urban hydrologist Murray McPherson was emphatically pushing for planning of the water economy to meet social and ecological requirements as early as 1970.

There has been a failure to take precautions when dealing with water in the past. Problems arising from excessive consumption were often not recognized in time. And then when the problems were recognized they did not all generate appropriate pressure leading to political action, not all the solutions that were determined politically led to decisions that could be implemented, and those decisions did not all lead to concrete measures. Such measures were frequently consequence-driven, local case-by-case decisions that were made in response to damage, but not to causes. Here the 'enemy approach' was generally taken: excess or dirty water had to be removed from towns as quickly as possible. Measures were designed to meet a purpose, and not integrated into comprehensive planning appropriate to the complexity of the water cycle. Thus the groundwater level was inevitably lowered in many urban areas, flooding increased and natural plant and animal habitats were destroyed. The larger cities become, the more they seem to use water regardless of the consequences. For example, Peking is a city with millions of inhabitants. The groundwater level is going down annually by over 2 metres, but water is used for air conditioning plants, cleaning cars and street cleaning, huge sprinkler systems are installed for green areas and rainwater is removed from the city in large channels. A Mediterranean tourist uses a thousand litres of water a day, even though it is a particular scarce commodity in the region in the summer months. Water is wasted all over the world, in countries with rapidly growing cities that are in the early stages of industrialization, in industrialized countries growing at a moderate rate, in regions that have little water and regions that have a lot of water. At the same time there are already a billion people who do not have adequate supplies of drinking water, two billion people have no sanitary facilities and four billion people produce contaminated water that is not subsequently purified to a sufficient extent.

Water management can only be balanced if social and economic wishes are covered by the quantities of available and renewable water. We have to accept that urban water concepts cannot be based on prefabricated models, whether they are local or imported. Water problems must be solved specifically and within the immediate vicinity for every town, every district and even every neighbourhood. Something that works for a town can be inappropriate in a particular neighbourhood. Realizing this compels us to decentralize responsibility and action. There are many reasons for this. Large supply and disposal systems cost far more than small, autonomous systems. Small units are far less prone to faults. They bring small and middle-sized enterprises together to construct and maintain them, and thus reinforce socio-economic structures. Small, autonomous systems remain able to function because the people running and using them identify with their system and see it as their property. 'Water neighbourhoods' are also better able to take responsibility for preventive measures.

So solving water problems in town requires a dual system. Technically speaking, local resources have to be used. Rainwater management is the key to the future. Water is part of a cycle in which it is used to water green areas and feed ponds that enhance the value of the immediate environment. Local people take the initiative in small-scale water-neighbourhoods. Public water supplies can then be reduced to covering a basic load, dependent on local climatic conditions. Economic responsibility is taken for a small area: the water neighbourhoods have to buy water in from central suppliers. In terms of water prices, a clear distinction has to be made between value, costs and tariffs. Here the real cost of supply and disposal has to be met by the user.

This system has a chance of success in the mega-cities because of the living conditions there: people live in a confined local environment, and spend their evenings and weekends in the immediate vicinity. They rely on local shopping facilities and leisure activities. Thus as a rule town-dwellers lead a life that is restricted to the locality, regardless of the size of their town. So the city of the future will have to be a city of neighbourhoods in which life-style and development are determined on a small scale. The central water authorities then follow the wholesale principle and sell to the neighbourhood units, who then manage their water internally.

The question remains of how long it will take to rethink in this way. Hesiod established the basic link between water pollution and the health of townspeople as early as 800 BC. At that time it took about three centuries for the Greek cities to introduce sanitary installation of the kind that already existed in the early cultures of Mesopotamia and on the Indus. We are faced with a learning process that will start in school and continue throughout our lifetimes. Whatever happens, this new way of dealing with water can only come from the inside, from the user.

As eco-systems do not respect national boundaries, internationally agreed water management is essential. But within this global network the regulation systems must leave sufficient scope for regional and local implementation. It would be wrong to see globalization as doing everything in the same way. The basic principles must be recognized globally and implemented by regulation – then the appropriate solutions have to be found locally.

Global action should lead to solidarity in dealing with water. Here responsibility still lies with the industrialized countries. They are in a good position economically, and must therefore begin to implement the new thinking, particularly as they have dumped the cost of their growth on to nature and the environment in the past. Here making pretty declarations of intent about water protection is just as inadequate as suggesting to developing countries that they should handle their resources carefully. The development and environmental crises that the industrialized countries went through in the eighties have still not been fully overcome.

Global economic competition between cities must be transformed into global competition for the best ecological conditions, which will make cities economically competitive in the long term again.

放眼全球·本土行动

沃夫冈·F·盖格教授

水既不是取之不尽、用之不竭的，也不是无懈可击的。但如今，对于它的使用强度却往往忽略了这些事实，我们正在越来越多地利用和污染这个大自然恩赐人类的、对于生存必不可缺的礼物。如果我们不希望在未来无水可用，我们必须协同思考，下放权力，在地方一级建立水资源利用自治系统。

没有哪种自然资源能够与人类对于水资源的繁重、复杂需求相比。即使它部分地不可再生，但与其他资源相比，我们却大大地将它忽略，就如人们只记得如何进行国际化协调开采石油资源。我们对待水资源的态度就好像它是取之不尽、用之不竭的。从哲学、科学和技术层面，都证实了这个错误的评估。

整体而言，人们喜欢并总是在靠近水源的地方建立城镇，这样它就可以直接被使用。它还作为传输的媒介，对于促进贸易和为居民的生活带来福祉做出贡献。城镇之中的水体具备文化、建筑和社会功能。早在1970年，城市水文学家穆雷·麦弗逊即强调了运用水经济进行规划，以满足社会和生态需求的观点。

过去在处理水资源时所采取的预防措施已被证明失败。过度消费中出现的问题在当时往往被忽略掉。当问题显现的时候，它们也并没有激发应有的、能够产生政治行动的压力，且并非所有的解决方案能够从政治层面上可以解决实际的问题，这些决定没有全部成为解决问题的具体措施。所以，这些措施通常更关注于结果，而地方层面上逐案解决的方式仅关注于危害本身，没有对其产生的原因采取本质的治理。结果，"消灭敌人的做法"就被普遍地采用，多余的或脏水需要被马上移出城镇。所采用的措施仅仅为了单一的目的，而非考虑水循环的复杂性而被纳入到全面的规划之中。如此这般，许多城市地区的地下水位不可避免地降低，洪灾发生的可能性增大，天然的动植物栖息环境被破坏。

城市规模变得越大，它们不顾后果地使用水资源的状况变得越明显。例如，北京是一个拥有百万人口规模的大都市。它的地下水位以每年超过2m的速度递减，而与此同时，城市水资源被广泛使用于空调厂、车辆和街道清洁、绿地区域内安装的众多喷水系统，而降雨却被收集在城市巨大的地下管道之中被排放到城市之外。一个来到地中海的游客每天使用1000升水，即便它是当地夏季数月之中的一种尤为稀缺的商品。不论是在经济快速增长的发展中国家、经济发展平稳的发达国家，水资源短缺的地区以及拥有大量水体的地区，水资源都在被浪费。与此同时，已经有10亿人得不到足够的饮用水供应，20亿人享受不到合格的卫生设施，40亿人口制造着废水，这些废水在产生之后并没有被净化达到相应的程度。

水资源管理只有在社会和经济的愿望真正将水资源的可得到和可再生数量纳入考虑的情况下，才能够取得平衡。我们必须接受，城市的水管理概念不可能建立在无论是本地或是引入的预制模型的基础上。有关水的问题必须在每一个具体的城市、区域，甚至每一个邻里之间，进行针对性地解决。在一个小镇行之有效的做法被应用在它的毗邻地区就可能是不合适的。意识到这一点，便迫使我们将责任和行动权力下放。这样做的原因有很多。大型供水和污水处理系统的成本远远高于小型的自治系统。小型的设备不太容易出现故障。它们带来了中、小规模的企业共同建设和维护，从而稳固了社会经济结构。小型自治系统能够发挥正常的功能，因为人们将之融入自己所有的资产对其进行使用和运行，并把它们看作是自己的财产。"水邻居"因此也能够更好地承担预防措施的责任。

所以，解决城镇水问题需要双重系统。技术层面上，需要使用本地资源。雨水管理是未来发展的关键。水是循环的一部分，在这其中，水被用来浇灌绿地、补给池塘用水，这些都对于提升直接的环境价值发挥了作用。当地群众对于小型的人水邻里关系具有较大的主动性。因此，依赖于当地的气候条件，公共性供水覆盖可以适当减少至仅覆盖基本需求。经济责任被下放至一个小的范围：水社区需要向中央的供应商购买水资源。在水价方面，必须在价值、成本和税收方面做出明确划分。供水和污水处理的实际成本需要获得用户的认可。

该系统在特大城市有可能获得成功，因为那里具备如下的生存条件：人们生活在一个特定的本地环境，晚上和周末时光都在附近区域渡过。他们依赖当地的购物设施和休闲活动场所。因此，作为一种规则，不管城镇的规模大小，城镇居民的生活基本上被限制于当地的环境之中。因此，未来的城市将是一个邻里社区城市，生活方式及其发展会在小规模范围内进行。中央水行政主管部门需要遵照批发的原则，出售给社区单位，它们可以在其内部进行用水管理。

或许仍然存在疑问，以这样的方式进行重新思考需要花费多长的时间。早在公元前800年，赫西奥德即提出水污染与城镇居民的健康之间的关联。在当时，希腊城市经历了大约三个世纪来推介早已在印度河流域的美索不达米亚早期文化之中存在的卫生洁具的安装。我们正处于一个学习的过程，它从学校之中开始，并将持续我们的一生。无论发生什么事情，这种新的水处理方式只能来自内部，来自使用者本身的体会、感悟和总结。

生态系统无国界，所以跨国进行的水管理是必要的。但是在这个全球网络中，监管系统必须为区域和地方实施留出足够的空间。将全球化视为以同样的方式做每一件事情是错误的。其基本原则必须做到全球范围的认知，按照法规实施监管，而合适的解决方案只能在地方层面上发现和寻找得到。

全球的行动会带来在处理水问题时的团结。在这方面，责任仍然主要在于发达国家。他们处于经济上的优势地位，所以必须开始实施新的思维，特别是考虑他们将过去经济增长对于自然和环境造成的影响抛甩出来。这里，发表动听声明的意图即是为弥补为发展中国家在仔细认真地应对和处理自身资源问题方面建言献策的不足。工业化国家在20世纪80年代经历的发展和环境危机至今仍未被完全克服。

城市间的全球经济竞争必须转变为全球最佳生态环境的竞争，这将再次为城市带来长期的经济竞争力。

F011 Residential Projects 住宅项目

Solar City, Linz, Austria
奥地利林茨太阳城

At the beginning of the third millennium Solar City is Austria's largest urban development project. This first phase including a neighborhood of 4,500 residents is designed for a site of 32.5ha, including 20ha of open space. The vision is for totally sustainable architecture and urban living, creating a new benchmark in high quality city design.

Atelier Dreiseitl worked together with renowned Architects - Norman Forster, Richard Rogers, Thomas Herzog to compose an overall sustainable urban concept. The innovative landscape concepts offers outstanding recreational spaces within the urban area, including a new beach and swimming lake, thus protecting the adjacent Traun-Danube wetland woodland, a registered "Natura 2000" site. Stormwater management inside of the new housing development helps suppor t the water table and water quality in the floodplain and serves to increase the local aquifer recharge inside of the built environment. The surface drainage is handled with open channels and swales and infiltrates mostly inside the green belt. Design guidelines were developed to facilitate the long term implementation of the design concepts into a phased construction process.

在新千年的开始，太阳城成为了奥地利最大的城市开发项目。包括4500名居民社区的一期设计阶段占地面积32.5公顷，包含20公顷的开放空间面积。项目的愿景为建造完全可持续性的建筑和城市生活空间，并成为高品质城市设计的新标杆。

德国戴水道设计公司与几位著名的建筑师——诺曼·福斯特、理查德·罗杰斯、托马斯·赫尔佐格合作，设计出一个整体的可持续发展城市概念。创新的景观概念在市区内提供出色的休闲空间，包括一个新海滩和可进入游泳的湖体，从而保护邻近的、已被注册载入"自然2000"保护区域的特劳恩-多瑙河湿地林地。新住宅开发项目中的雨水管理计划，能够有效地保持河滩区域的水位线和水质，并增加建筑环境内部的含水层补给。地面排水由大部分位于绿带之中的开敞式沟渠和洼地进行处理、过滤。设计原则即为方便对于长期的实施理念进行分阶段的建造施工。

Location Linz, Austria　项目地址 奥地利 林茨
Client City of Linz　项目委托 林茨市政府
Urban Plan READ Group　规划设计 READ Group
Area 32.5ha　项目面积 32.5公顷
Construction 2005　建成时间 2005年
Awards UN Best Practice Award　所获奖项 联合国最佳实践奖

1 The new Solar City at the Danube river
多瑙河岸边的新建太阳城

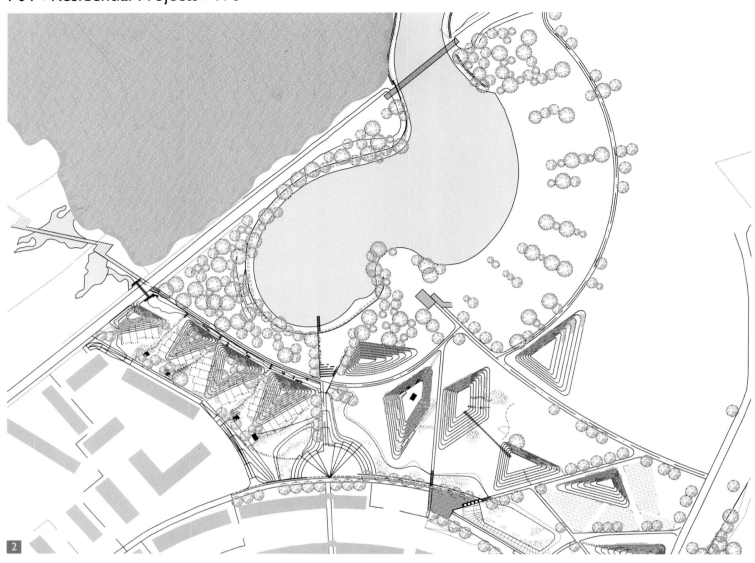

2 Landscape plan with the new lake
具有新湖体特征的景观规划
3 Lake as the main recreational element
湖体成为主要的休闲娱乐元素
4 Playground on the landscape forms
景观形式上的游乐场地
5 Plazas with shading structure
拥有遮荫结构的广场
6 Bioswales integrated in urban design
整合生态洼地至城市设计之中
7 Youth testing their boldness
年轻人展示他们的强健体魄

F02 | Residential Projects 住宅项目

Zhangjiawo Neighbourhood Community, Tianjin
天津张家窝新镇社会山小区

Location Tianjin, China
Client Shanghai Wisepool Real Estate Co., Ltd.
Landscape Architect Atelier Dreiseitl
Architect Schaller/Theodor Architekten, Schmitz Architekten, WLA
Area Phase 1: 20 hectares, Master Plan: 180 hectares
Construction Period 2006~2009
Completion Time 2009
项目地址 中国 天津
项目委托 上海国民实业有限公司
景观设计 德国戴水道设计公司
建筑设计 Schaller/Theodor Architekten, Schmitz Architekten, WLA
项目面积 总体规划180公顷，其中一期规划面积20公顷
施工时间 2006年~2009年
建成时间 2009年

The need for new housing in China reflects on one hand an expanding urban population and on the other an important shift in expectations. The developer chose a design team who were to deliver a new urban housing concept specific to the site and of outstanding "liveability". The first 20 hectares have been built and reveal the design qualities at the heart of the development concept.

Delivering a new urban town masterplan of notable outstanding 'livability,' the first 20 hectares have been built and reveals the core qualities of our urban design concept. Subdivided into high-density communities, pedestrian paths were interlaced throughout and connected to major circulation arteries and public buildings. Excellent public transportation allowed parking to be minimized and the space used for intimate green spaces, courtyards and plazas. These spaces form a soft stormwater mangement infrastructure, infiltrating rainwater on site and supplying the adjacent restored river with clean water. Existing orchard trees, some over 200 years old were retained. The 30 meter wide irrigation canal was restored as a river, spanning the length of 1 kilometer. Accessibility by boardwalks, steps and ramps, the restored ecological integrity of the rivercanal has created a community treasure and a focus point for leisure and socializing.

在中国，新住房的需求一方面反映了不断膨胀的城市人口，另一方面反映人们对于住房预期的重要转变。开发商所选择的设计团队采用了与场地非常契合的新城市居住理念，并融合了出色的"宜居性"。一期20公顷的区域已建设完成，彰显出设计质量在开发理念中的核心地位。

戴水道设计团队采用与场地契合的新城市居住理念，并融合出色的"宜居性"设计，首期20公顷的区域已建设完成，彰显出设计质量在开发理念中的核心地位。在高密度社区之中，步行通道网络遍布，与主要交通干道和公共建筑直接连通。便利的公共交通将私人车行数量最小化，闲下的场地空间则形成各类绿地、院落和广场。它们共同组成软质的雨水管理基础设施，将场地内雨水过滤，为临近修复河流提供洁净的水源。场地中原有的果树林木被保留下来，其中一些树龄已经超过200年，它们被融入整体的设计之中。丰产河原本宽30米的污水沟被修复成为长1000米的美丽、洁净的自然河流，它缓缓蜿蜒，周围植被密布，在遭受了多年工农业污染之后得到修复。通过木栈道、台阶和坡道的连接，经生态修复、整合的丰产河成为了社区的财富以及休闲娱乐、社交活动的中心区域。

1 Internal watercanal as a park
内部水体组成的社区雨洪公园

2 Masterplan with the river
   总体规划及河流
3 Internal water design
   内部水系设计
4 Club house with water plaza
   水广场会所
5-9 Lujuba trees as strong characters
   枣树成为了场地明显的特征

住宅项目 | 181

10 Central water park of phase 2
二期中央水园
11 Light shading structure
遮光结构
12 Masterplan
规划图
13 Wave deck for relaxing
休闲波形平板

F03 | Residential Projects 住宅项目

A Climate Adaptation Community, Arkadien Winnenden, Germany
德国阿卡迪亚温嫩登气候适应型社区

Arkadien Winnenden is a hardcore industrial regeneration project. A diversity of high performance components make this the world's most sustainable neighbourhood and provides a fresh new vision for people-friendly and resource productive suburbs.

Mixed architectural typologies are kept a cohesive neighbourhood thanks to the appealing Mediterranean colour concept and "garden city" quality of the streetscapes. Water sensitive urban design provides a distinctive urban character. Street corners are mini-plazas and places to chat with your neighbour or for kids to kick a ball. Although the streetscapes are distinctively pedestrian, a shared circulation concept means that the site is fully accessible for vehicles, with parking options in an underground garage, carports, and parking spots neatly tucked away between gardens on the unique load-bearing planting substrate.

阿卡迪亚温嫩登是一个核心的工业再生项目。多样化的高性能组成部分将此社区打造成为世界上最可持续发展的社区，并提供了一个拥有全新视角的人类友好、资源有效型的郊区典范。

混合的建筑类型，结合地中海地区迷人的色彩理念和符合"花园城市"质量标准的美丽街景，形成了一个拥有凝聚力的社区。水敏感城市设计为此项目提供了一个独特的城市性格。街角的小型广场、空间可以为邻里交谈、孩子们踢球玩耍提供场地。虽然街景专为步行而准备，共享的交通概念同时意味着车辆也完全可以进入该场地，而地下车库、停车棚、以及隐藏于独特的、能够承担负重的种植基质的花园之间、整齐排布的停车场地，为车辆的停放提供了多样的选择。

Location Winnenden, Germany
Client Strenger Bau und Wohnen GmbH
Partner Eble Architecture
Area 34,000m²
Completion Time 2012.3
Awards Green Dot Award 2011; International Project Award for Livable Communities Finalist 2012; World Architecture Network Urban Regeneration Award Finalist 2012

项目地址 德国 温嫩登
项目委托 Strenger Bau und Wohnen GmbH
合作机构 Eble Architecture
项目面积 34000平方米
建成时间 2012年3月
所获奖项 2011绿点奖、
2012国际宜居社区项目入围奖、
2012世界建筑网络城市再生项目入围奖

1　Central Lake for the community
　　社区内中心湖体

2   Playground integrated as flood plain
    整合了泛洪场地的游乐区域
3,8 Flood protection impacts the entire Rhein
    作用于整个莱茵河沿岸的洪水保护措施
4-7 Open space designed for community activities
    为社区活动进行的开放空间设计

住宅项目 | 187

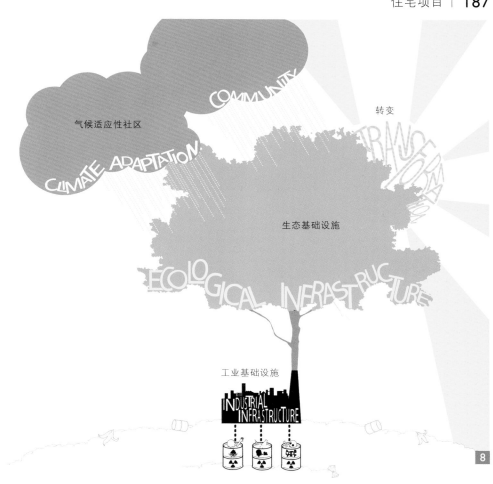

气候适应性社区　　转变

生态基础设施

工业基础设施

F03 | Residential Projects | 188

Ocean 海洋

9

River 河流　　Lake 湖泊　　Cleansing 净化作用

9 Stormwater tools are integrated
  整合的雨洪控制方法
10 Human scale areas attract all
  人性化尺度设计具有吸引力
11-13 Shared space works for all
  众人皆可使用的公共空间

Guorui International Investment Square, Beijing, China
中国北京国锐国际投资广场

Location Beijing, China
Client Beijing Guorui Investment
Partner Foster&Partners
Area 90,000m²
Design Time 2010-2012
Construction 2012-ongoing
Completion Time expected in 2014
项目地址 中国 北京
项目委托 北京国锐投资有限公司
合作机构 Foster&Partners
项目面积 90000平方米
设计时间 2010年~2012年
施工时间 2012年至今
建成时间 预计2014年

Guo Rui Square is a new mixed use development in Beijing located in the rapid developing international district of Beijing Yizhuang Economic-Technological Development Area (BDA). Atelier Dreiseitl developed in close collaboration with lead architect Fosters and Partners the project aims to push the boundaries of mixed use developments.

The design demonstrates how a sustainable use of water can be embedded in contemporary landscape architecture and creates vibrant urban areas as well as a personal environment rich in experiences.

国锐广场坐落于经济快速发展的北京亦庄经济技术开发区内，是一个新建、多用途开发项目。德国戴水道设计公司与建筑行业领先的福斯特事务所合作开发，该项目旨在推动多用途开发项目的新模式。

此设计展示了水的可持续利用如何与当代景观设计有效地结合，创造充满活力的城市区域环境并丰富个人的环境体验。

1 Dense community with a central park
   高密度社区内部拥有中心公园
2 Water table, walking boulevard in between high-rise dwelling buildings supply a livable setting
   水景台、步行林荫路等景观设计提供了高层密集住宅之中的宜居环境

F04 | Residential Projects | 192

Stormwater Concept:
1. Public sewer
2. Bio-swale with infiltration and retention
3. Pretreatment and sedimentation
4. Reduced outflow
5. Green and permeable pavement
6. Irrigation
7. Refill
8. Green roofs
9. Reflecting pool
10. Stormwater run off
11. Treatment room
12. Underground storage for stormwater detention and reuse
13. Cleansing biotope
14. Bio-swale with infiltration and retention

雨洪利用概念图：
1. 市政雨水管道
2. 渗透和滞留洼地
3. 预处理和沉淀
4. 削峰外排
5. 绿地和可渗透铺装
6. 灌溉
7. 补湖
8. 屋顶花园
9. 镜面水池
10. 径流收集
11. 净化处理间
12. 地下储水设备用于雨水滞留和收集利用
13. 生态净化群落
14. 渗透和滞留洼地

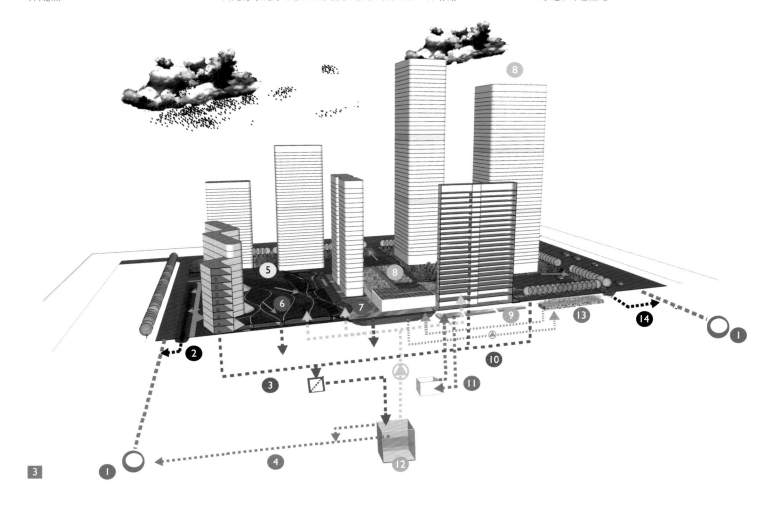

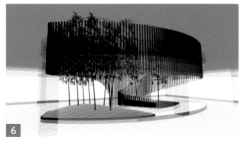

3　Water management saves money
　　雨洪管理能够节约成本
4-6　Public plaza as a gate to the city
　　公共广场成为与城市连接的门户
7　Balance of open meadows and trees for shade
　　开阔的花草地与树影婆娑相得益彰
8　Internal park with underground parking
　　设有地下停车场的社区内部公园

Located in the heart of the site is a lake which is supplied by an underground stormwater system utilizing rainwater collected from the building roofs and within the park. Still Water Basins reflect the architecture and the surrounding. Water Jets animate public plazas areas while Cleansing Biotops maintain of the high water quality and contribute to the overall outdoor comfort.

The rolling park landscape for the residential park provides a large variety of activity areas for its residents with space for individual recreation as well as group activities providing ever-changing vistas through the undulating park topography.

The public areas provide a high quality setting for urban life; discovering artwork and water tables whilst strolling along promenades, gathering underneath tree canopies for a chat or participating in large scale events on the public plazas.

场地的中心拥有湖体，由一个地下雨水管理系统收集来自屋顶以及公园地面上的雨水，提供水源。静水平台的设计反射建筑及周围环境结构的影像；水景喷泉激活公共广场空间；而生态净化群落维护了高标准的水质，并增强了整体室外环境的舒适度。

为住宅园区设计的流动公园景观给身住其中的居民们提供了种类繁多的个体休闲以及群体活动空间，通过地形起伏提供不断变化的园林景致。

公共区域提供了高品质的城市生活空间，可以在闲庭信步、三五成群地在树荫下乘凉聊天、或是与公共广场上参加大型活动的同时，探索并欣赏艺术和水景景观。

Art as an Icebreaker
by Herbert Dreiseitl

Every artist and planner realizes when designing a water feature that water needs a great deal of respect when you are working with it: it is mysterious, and likes for be investigated sensitively. Thorough observation of water in a landscape is particularly valuable. We are constantly discovering that water is our best teacher. Despite all prior considerations and even with a great deal of experience: water does not always behave as expected.

Corrections and constant new experiments are the principal task. The designs improve gradually and start to carry the signature of the water itself. Water influences our sense of well-being in towns and in buildings, it affects the humidity, the temperature, the cleanness of the air, the climate. Water can be used in such a way that it filters, cools or warms the outside air and regulates humidity. The sounds of water can also be against street noise. They are smoothing, and compensate for urban stress, moving light to life. Water distorts, refracts and reflects. It distributes incident light dynamically in connection with its own movement. And finally, water creates atmosphere, something that is vital to our towns and cities if they are to be individual, unmistakable and easily recognized, with a sense of being home. This is a difficult set of phenomena to describe, and has something to do with the spiritual quality of a place, defining life and movement, which is something that water can convey directly like no other element.

Today it is scarcely possible to understand any longer how closely the fields of art, architecture and engineering were linked until the late Middle Ages: they formed a unit. An outstanding example of this kind of interdisciplinary work with water is an artist like Leonardo da Vinci. However, we are becoming increasingly aware of the necessity to work sustainably and far-sightedly with water. In the future planners will be increasingly required to put water itself back in a consistent context. This is the only appropriate way for dealing with its qualities and diversity. Water always creates a relationship between detail and the whole. Each individual drop contributes to the balance of the earth's climate. Water projects become valuable when they help this process and can show that the place is being addressed, and how it is connected with the world around it.

Water projects should include citizens and later users in planning and decision-making to as large an extent as possible. It is important to promote and stimulate people's own creativity, as water is full of imagination. A water-playground is one of the most popular places there are, and not just for children.

让艺术成为设计的破冰者

赫伯特·德赖赛特尔

每一个艺术家和设计师都意识到,在面对水景设计的过程中需要给予足够的重视:水元素神秘变幻,需要对其进行细致入微的调研。对于景观之中水资源状况进行全面透彻的观察则显得尤为可贵。我们不断地发现,水是我们最好的老师。有时,即便进行了前期的思考和准备,并拥有大量的经验,它也并不总是如人预期般地发展。

完善和不断推陈出新的实验项目成为了解水体特征的主要方式。设计逐步改善并开始顺应水体自身的特征。水影响着我们在城镇和住宅之中的幸福感受,它影响着湿度、温度、空气清洁度,甚至整体气候。水可以过滤、冷却或加热外部空气温度,并调节空气湿度。水声还能够屏蔽街道的噪音。它们是柔和的,能够缓和城市压力,为城市空间带来亮色。水体形态随意变幻,能够折射和反射光影。它能够以其自身的流动性反映光影的动态变幻。而如果作为单独存在的、表达准确且易于识别的景观构成,水则可以创建一种氛围,一些能为我们的城市带来活力的东西,为城市之中的居民带来如家般的感受。这很难用语言形容,它可以为场地精神品质的提升发挥作用,它还可以定义日常生活以及人们行为的方式,这些都是除却水元素之外的其他任何设计元素无法直接传递的。

现在的我们已经很难想象,在中世纪晚期之前,艺术、建筑和工程等领域曾经是多么紧密地联系在一起:它们形成了一个组合体。创作这种拥有水元素的跨学科设计作品的突出代表人物即为达芬奇等艺术家。如今,我们正在越来越多地意识到可持续、富有远见地开发水资源的必要性。今后,规划者们需要更多地将水体恢复至其一贯存在的背景之中。这也是反映它的品质和多样变化的唯一恰当方式。水总是能够创建细节与整体相互关联的关系。每一滴水珠都对地球气候的平衡发挥作用。因此,与此过程相关的水景项目就变得有价值,能够显示该场地的定位以及与周围世界的联系。

在与水相关的项目规划和决策阶段应该尽可能多地吸收和采纳市民和使用者的意见。他们的投入和参与是非常重要的,水体是充满想象的,它可以促进和激发人们的自身创造力。嬉水游乐园对于儿童乃至更大范围的人群来讲都是最受欢迎的场所之一。

Water playground BUGA, Koblenz, Germany
德国科布伦茨BUGA嬉水游乐场

Location Koblenz, Germany
Client BUGA Koblenz 2011 GmbH
Area 1,111m²
Completion Time 2011
项目地址 德国 科布伦茨
项目委托 2011科布伦茨BUGA联邦园艺展
项目面积 1111平方米
建成时间 2011年

Water is ubiquitous at Deutsches Eck in Koblenz where the two rivers, the Rhine and the Mosel, meet and join to one. We are surrounded by water when standing at this impressive and historic landmark. But is it possible to physically interact with the water here? Through the expanded and renovated riverside promenade, the rivers are close, but in order to live up to the motto of the Bundesgar tenschau "Koblenz transforms", the water has to become accessible, touchable and tangible. In our view, this experience should be realized with the water playground which is situated only a few steps away from the Deutsche Eck. This water playground will be one of the four theme playgrounds of the BUGA. The starting point of our work was to develop a place to experience water, incorporating the theme of the two rivers, the Rhine and the Mosel. It was thought to present different landscape and water characteristics of the rivers and hereby work on the differences between the two rivers. In addition, beside the permanent construction of this water playground, a temporary exhibition ship would display the work of the water federation and ship administration to visitors.

位于德国科布伦茨市莱茵河和摩泽尔河两条河流交汇处的"德意志之角"，水源充沛。在这个水流环绕之地，同时拥有着令人印象深刻、具有历史意义的标志性建筑。在如此的场地，是否能够亲身体会与水的互动？行走在扩建、改建后的滨江大道之上，河流看似与人们很接近，但为了符合作为"科布伦茨转型"契机之德联邦BUGA园艺展的要求，水体应成为可进入、可被触知、且现实有形的。这样的体验应该通过距离"德意志之角"几步之遥的水上游乐场得以实现。它将作为BUGA联邦园艺展的四个主题游乐场之一。设计的出发点即为结合莱茵河和摩泽尔河，开发能够进行亲水体验的场地。通过呈现河流景观和水特点的差异，进而开发两条河流之不同。此外，水游乐场永久性建筑旁边的一处临时展示舰，将作为水联合会和船舶管理委员会的展示平台，向游人开放。

1-3 Water playing in a smooth topography
在平滑的地形上嬉水

Taipei Public Art Project – Water-art Red Snake
中国台北公共艺术设计——红蛇水景艺术装置

Location Taipei City
Client City of Taipei
Partner EDS International
Area 840m²

项目地址 中国台北市
项目委托 中国台北市
合作机构 EDS International
项目面积 840平方米

This proposal makes a step forward to enhance the experience of water along a City River in Taipei. Its main purpose was to evoke water's spirit of change and to invite people to interact with an art piece that connects ideas about the city's background, culture, conservation and the environment. The public art work was submitted as part of a competition for the improvement of the Chung-Gone drainage system & Corridor Environment in Taipei County.

Measuring up to 40m, the concept of the artwork is to extend and connect people's emotions and feeling by experiencing the different facets of water. Made of shiny red metallic, the sculpture shows flowing water forms that move through the urban landscape in an interactive artwork manner. Walking along side the artwork can be described as walking along the same journey that water makes in an articulated landscape. By communicating this experience the public gets to enjoy 4 different experiences of water flows as articulated by the art sculpture. Like the warm morning sun, mist will give a welcoming experience, providing the feeling of an early morning breeze and lightness to people, in contrast to the hard and bustling life that is typical in a dense city. To enjoy a moment of the experience of water droplets touching our skin brings us closer to experience as moment of cleansing that leaves us rejuvenated, cool and refreshed.

Always in motion, the flowing nature of water shares a sense of movement and never ending energetic force that creates both dynamism and changes, which reflects a large extension the society that we live in. The opposite of motion, water tranquility provides people a chance to take a short break and reflect on the wonders that nature has provided for us and that we should be grateful for.

这项水景艺术装置提案为提升市民沿台北城市河流的亲水体验方面向前迈进了一步。其主要目的是唤起水体流动变幻的精髓并邀请人们与此连接城市的背景、文化、环境保护等观念的艺术装置进行互动。这项公共艺术作品即作为一个旨在提升、改善台北县仲刚排水系统及廊道环境的竞赛项目的一部分。

高达40米，该艺术作品的设计理念为通过亲水体验的不同侧面，扩展并连接人们的情绪和感受。使用闪亮的红色金属材料，该雕塑作品通过互动的艺术形式，展示了在城市景观当中游走的流水形态。沿着该艺术品的行走过程，就正如水流在连接的景观之中流动一般。通过这样的交流体验，市民就可以随着艺术雕塑作品的连接而感受到四种不同的水体流动体验。就如温暖的清晨阳光、雾气将给予人们幻影般的体验，为人们提供清晨微风般、明亮的感觉，与典型的人口密集的繁华都市之中辛苦、熙熙攘攘的生活形成反差。享受片刻的水滴接触肌肤之感，即可带给我们与水更为接近的时刻，让我们焕发青春、感觉清新、凉爽。

处于动态变化之中，水流永不停歇的自然属性让我们分享了其运动感以及永不停止、充满活力的力量，创造出活力和变化，反映了我们所生活的社会巨大的扩展性。而与水流的运动相反的是水的安宁，它为人们提供片刻休息的机会、给予我们所应该懂得感激的自然奇迹。

1-3 Riverfront with a breathtaking water tube
河滨区域非凡的艺术水管设计

Art Wall, Tanner Springs Park, Portland, USA
美国波特兰坦纳斯普林斯公园艺术墙

Location Portland, USA
Client Portland
Partner GreenWorks PC
Completion Time 2005
Awards
ULI Open Space Award Finalist 2011;
ASLA Oregon Chapter, Merit Award Landscape Design 2006

项目地址 美国 波特兰
项目委托 波特兰市政府
合作机构 GreenWorks PC
建成时间 2005年
所获奖项
2011年城市土地学会开放空间设计奖入围；
2006年美国景观设计师协会俄勒冈州景观设计优胜奖

A floating pontoon crosses the water, literally swimming across its invisible and varying depths. Symbolic of the old city fabric, historic railroad tracks form a wave-wall along one side of the pond. Called the 'Art Wall', there is a harmonious contrast between the static strength of the rail tracks and the lithe and flowing movement of the wall as it oscillates in and out, the top also rising and falling. The verticality of the rail tracks is surprisingly filigree. The movement of the wall is doubled in the reflection in the water.

The Art Wall is 60 meters' long and composed of 368 rails. 99 pieces of fused glass are inset with images of dragonflies, spiders, amphibians and insects, like animals captured in amber, creatures of times and habitats long gone. The images were hand-painted by artist Herbert Dreiseitl directly on Portland glass, which was then fused and melted to achieve the final effect.

一座浮桥穿越水面，在无形、变换深度的水面之上游弋穿行。作为老城区组成结构的象征，具有历史意义的铁轨组成了一道沿池塘边壁的波浪状墙体。"艺术墙"以其形式的凸凹变化、顶部的升降变幻，成为了铁轨静态坚硬造型以及墙体轻盈、流畅流动形式的和谐对比。铁轨的垂直特征极具装饰性。墙体的流动在水体的反射下成双。

艺术墙长60米，由368条铁轨组成。99件熔融玻璃内嵌入蜻蜓、蜘蛛、两栖类以及昆虫的图像，如琥珀化石之中的动物一样，收集了时光久远、生境绝迹的生物形象。图像由艺术家赫伯特·德赖赛特尔先生直接手绘于波特兰的玻璃之上，并进行熔融以达到最终的效果。

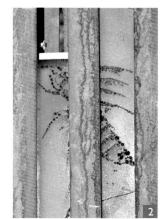

1-3 Undulating art wall out of rails
起伏的铁轨艺术墙体设计

Central Fountain, Ritz Carlton, Tianjin Taiandao, China
中国天津泰安道丽思卡尔顿酒店中央水景

Location Tianjin, China
Client Tianjin Tiantai Real Estate Development Co., Ltd.
Area 8,800m²
Design Time 2011-2012
Construction 2013
Completion Time 2013
项目地址 中国 天津
项目委托 天津市天泰置业发展有限公司
项目面积 8800平方米
设计时间 2011~2012年
施工时间 2013年
建成时间 2013年

Tianjin's new Ritz Carlton Hotel is located in the former British concession of Tianjin, in the heart of the city. The hotel's classical architecture with its courtyard is responding to the areas historic background in the rapid pace of Tianjin's modern development. The courtyard is a modern interpretation of an English garden with a water feature, derived from the shape of a Victorian piece of jewelry, located in its center.

天津的新丽思卡尔顿酒店位于天津原英租界旧址——城市的中心地带。附有庭院的酒店古典建筑形式,反映了在快速发展的天津现代化发展步伐之中该地区的历史背景。中心庭院是一座英式水景花园的现代阐释。在其中央位置一个形如维多利亚时代的首饰之中,水体流淌而出。

1. Sound and aesthetics create special atmosphere
   声效和美学创造出一处独特的环境氛围
2. Landscape in the courtyard
   庭院景观
3. Roof garden supplies a quiet and relax place for a break or meeting with friends
   屋顶花园提供了一处安静、悠闲的场地,可在此休闲或约见朋友

Light and Water installation, Breiteweg
布莱特威戈"光与水"的装置

Location Barleben, Germany
Client Municipality of Barleben
Partner Atelier David Fuchs
Planning and Design Time 2004-2006
Completion Time 2012
项目地址 德国 巴尔莱本
项目委托 巴尔莱本市政府
合作机构 Atelier David Fuchs
规划和设计时间 2004年~2006年
建成时间 2012年

The community Barleben is situated on a major intersection, 100km south west of Berlin. The place is dominated by a historic streetscape with a central square for market and festival activities. After the reunification of East and West Germany, the municipality experienced an economic boom. The innovative, independent and creative design of Breiteweg reflects the spirit of this vibrant and successful town and sets a benchmark for the quality of urban space. Once a village and then a crumbling lonely street, the "Breiteweg" has now become a centre of the community. In four groups, eight sculptures of water and light are installed as active and attractive areas of contrast in the long road space. The network structure made of steel mesh is stretched and directs the film of water. At night, the lights above shine and reflect on the flowing water. As it reaches the bottom, the water and light waves are continued, absorbed by the glass blocks in the road paving and extended forming a trail of light. This place is frequented by young and old, who come here to play or just relax. It is more than a playground and a light-water game. The installation gives Barleben a special atmosphere and creates a self-confidence that reflects the economic success and also sets an example for future cultural development.

巴尔莱本社区位于柏林西南100千米之外重要的交通交叉路口，以历史悠久的街景为特色，一处中央广场可举办各种集市和节庆活动。东、西德统一后，城市经历了经济的繁荣发展。创新、独立、富于创造性的"布莱特威戈"设计正是反映了这种充满活力、成功的城市精神，并被视为城市空间质量的一个基准。曾经的村庄和落寞街区，如今的"布莱特威戈"已然成为社区的中心。四个组团、八个"水与光"的雕塑已安装到位，它们作为漫长道路空间的对比，将此地改造为充满活力和吸引力的地区。钢丝网制成的网络结构被拉伸，引导水幕的流动。到了晚上，其上灯光闪耀，水光溢彩。当到达底部，水光波纹继续，融入路面铺装的玻璃模块之中，在灯光轨迹的伴随下一路延伸。年轻人和老年人经常来到这里游玩或仅是休闲、放松。它的功用已经超出仅仅作为一个操场和嬉水游乐场地的范畴。"光与水"装置的安装建造为巴尔莱本增添了一种非常独特的氛围，反映了区域经济繁荣发展所带来的自信心，同时也为未来文化的发展树立了一个榜样。

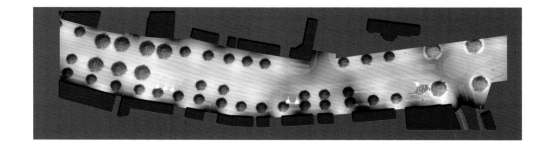

1　Water and light sculptures define focal points
　水和光影塑造场地焦点

Red Swings, Venlo, Netherlands
荷兰芬洛红色摇摆水景装置

Location Venlo, Netherlands
Client Floriade (An International Exhibition of Flowers and Gardening, Held Every 10 Years in the Netherlands)
Partner Atelier David Fuchs
Completion Time 2012
项目地址 荷兰 芬洛
项目委托 弗莱瑞雅德国际花卉园林展
合作机构 Atelier David Fuchs
建成时间 2012年

The over 4 hectares Floriadesee in Venlo/Netherlands will be certainly one of the highlights of the Floriade. Exclusively based on natural purification systems and only fed by rain water, it forms the framework for the World Show Stage, an event and exhibition area that is about art and culture from home and abroad. The Restaurant World-Show Stage is actually in the centre of the water. The water surrounding the restaurant area is about three feet higher than the rest of the main part of the lake. Here, water bubbles up like a source and constantly circulating water flows through a "delta" water play area in the lower lake. Through various dams and water wheels in this area, rivers and waterways can be played with and altered. The main attraction in this area is, however, the "swing"; an interactive object with an approximately 6.5m pipe and a flywheel, making it possible to draw water from the small lake and pass it over the delta in the main lake. The inside of the flywheel with a complex arrangement of gears, brake unit and counterweights also allows children to move the object. So, when the tube end is submerged in water, it is filled with about 50 liters of water. By rotating the flywheel, the pipe is pull up, whereby the water in the interior of the tube flows downwardly and gushes out through the bottom.

面积超过4公顷的荷兰芬洛弗莱瑞雅德国际花卉园林展无疑将成为城市的亮点之一。它完全基于自然净化系统、以雨水为源，组成了作为展示国内外艺术和文化活动的展览区域——"世界展示舞台"的框架结构。"世界展示舞台"餐厅是水体的中央，环绕餐厅区域的水体高于湖中其余主体部分约3英尺。在这里，仿佛水源般，水中气泡涌动，持续循环的水流通过"三角洲"水上游乐区进入地势较低的湖泊。通过该区域之中形式多样的水坝和水轮，河流和水道变得充满乐趣和变幻性。不过，这个区域最具吸引力的还是"红色摇摆"——一个拥有长约6.5米管道与飞轮的互动装置，通过它，小湖面之中的水拍打、流经"三角洲"，并汇入中心湖泊。飞轮的内部拥有复杂的齿轮、制动单元，其平衡装置能够允许儿童将其进行移动。因此，当管道端部浸没在水中，其内能够填充约50升的水量。通过旋转飞轮，管道被向上拉升，管道内部的水流从而向下流动，并通过底部涌出。

1-3 The swing, a new way of water play
轮摆成为了一种新型的嬉水方式

Water Curtain Fountain, Heiner-Metzger-Plaza, Neu-Ulm, Germany
德国新乌尔姆，海纳-梅茨格广场，水幕喷泉

Location Neu-Ulm, Germany
Client City of Neu-Ulm
Completion Time 2005
项目地址 德国 新乌尔姆
项目委托 新乌尔姆市政府
建成时间 2005年

The external corner, bounded by the two roads, is an ensemble of water screens. These are a highly visible identity marker, subtly and transparently shielding the plaza from traffic with suggestive views cut through. They define an inner space, which is active and fun with changing water patterns and intensities. The gurgling noise of the water creates a 'Soundscape'. Through another physical sense, the inclusiveness and fun atmosphere of the plaza is reinforced. Seating blocks were designed by students and sited together in the 5 centimeter shallow water screed.

广场的外部角落、两条街道的交叉处，坐落着一组整体水幕。这是一处非常明显的身份标志，通过建议性跨越的场景，巧妙、通透地将广场与城市交通屏蔽开来。水幕定义了一处内部空间，灵活、充满乐趣的改变着水流的方式和强度。潺潺的水流声创造了"声音景观"。通过这样一种外部感官体验，广场空间的包容性和欢乐气氛得到加强。座位模块由学生设计完成，被联排摆放于5厘米深的浅水之中。

1,3 Light water curtains-transparent or screening
薄薄的水帘清澈可见并可作为屏障之用
2,4 Children are enjoying the water feature
孩子们欢快地玩水嬉戏

Illustration Credits

CENTRAL WATERSHED MASTERPLAN, "ACTIVE BEAUTIFUL, CLEAN", SINGAPORE
新加坡中央地区水环境规划——"活力、美观、洁净"水敏城市设计导则
-1 ©Atelier Dreiseitl
-2 ©Dreiseitl/CH2MHill
-3,4,5,6,7,8,9,10,11,12,13,14 ©Atelier Dreiseitl

TIANJIN ECO CITY
天津生态城景观规划
-1 © Rheinschiene/Dreiseitl
-2 ©Atelier Dreiseitl
-3,4,5 ©Rheinschiene/Dreiseitl

TIANJIN HEXI CBD LANDSCAPE MASTER PLANNING
天津河西中央商务区整体景观规划
-1,2,3,4,5,6,7,8,9 ©Atelier Dreiseitl

DEVELOPMENT MASTERPLAN PINGDINGSHAN, HENAN, CHINA
河南平顶山生态景观与水文发展规划
-1,2,3,4,5 ©Pesch&Partner/Dreiseitl
-6,7,8 ©Atelier Dreiseitl

RESTORATION OF EMSCHER RIVER, RUHR VALLEY, GERMANY
德国埃姆舍河流域修复规划
-1, 2, 3 ©Atelier Dreiseitl

OFFENBACH HARBOUR, OFFENBACH, GERMANY
德国奥芬巴赫港口区域规划
-1,2 ©Atelier Dreiseitl
-3 ©Heidenreich
-4,5 ©Atelier Dreiseitl
-6,7 ©Heidenreich

ECOLOGICAL INFRASTRUCTURE & STORMWATER MANAGEMENT IN TIANJIN CULTURAL PARK
天津文化中心生态水环境基础设施
-1,2,3,4,5,6,7,8,9,10,12,13,14,15,16,18 ©Atelier Dreiseitl
-11 ©YMGC
- 17 ©Dreiseitl/Polyplan

WATER SYSTEM, POTSDAMER PLAZA, BERLIN, GERMANY
德国柏林波茨坦广场水系设计
-1,5 © J. Lee
-2,3,4, 6,7,8,9,10,11 ©Atelier Dreiseitl

TRANSFORMATION OF A CITY DISTRICT, TAIPEI, TAIWAN
台北城市街区的转变
-1 ©Eton Shu
-2,3,4,5,6,7,8,9,10,11,12 ©Atelier Dreiseitl

JTC, CLEAN TECH PARK, SINGAPORE
新加坡JTC清洁科技园
1,2,3,4,5,6,7,8 ©Atelier Dreiseitl

图片版权

KALLANG RIVERSIDE, SINGAPORE
新加坡加冷河滨区域规划
-1, ©Atelier Dreiseitl
-2 © Google Earth
-3 ©Dreiseitl/CPG
- 4,5,6,7,8,9,10 ©Atelier Dreiseitl

FENG CHAN RIVER RESTORATION, ZHANGJIAWO NEW TOWN, TIANJIN, CHINA
中国天津张家窝新镇，丰产河修复
-1,2,3,4,5,6,7,8 ©Atelier Dreiseitl

BISHAN-ANG MO KIO PARK AND KALLANG RIVER, SINGAPORE
新加坡碧山宏茂桥公园和加冷河修复
-1,3,4,5,6,8,11,12,13,14 ©Atelier Dreiseitl
-2 ©PUB & Dreiseitl
-7 ©Wang Guowen
-9 ©NParks
-10 ©PUB

ZOLLHALLEN PLAZA, FREIBURG, GERMANY
德国弗莱堡市扎哈伦广场
-1,2,4,9 © Karl Ludwig
-3,8,10 © Atelier Dreiseitl
- 5,6,7 © Doherty

DAYLIGHTING OF ALNA RIVER, HOLLALOKA, OSLO, NORWAY
挪威奥斯陆海伦拉卡，埃纳河的日光
-1,2,3,4,5,6,7 ©Atelier Dreiseitl

WATERTRACES, HANNOVERSCH MÜNDEN, GERMANY
德国汉恩蒙登"水之印"广场
-1,2,3,4,5,6,7 © Atelier Dreiseitl
-8 ©Herbert Dreiseitl

RESTORATION OF RIVER VOLME, HAGEN, GERMANY
德国哈根市沃尔姆河修复
 -1,2,3,4,5,6,7 ©Atelier Dreiseitl
-8 ©S Fuhrmann
-9 ©Atelier Dreiseitl

HEINER-METZGER-PLAZA, NEU-ULM, GERMANY
德国新乌尔姆，海纳-梅茨格广场
-1 ©Conne van D'Grachten
-2,3,4,5,6,7,8 © Atelier Dreiseitl

ROCHOR CANAL, SINGAPORE
新加坡梧槽运河修复
-1,2,3,4,5,6,7,8 ©Atelier Dreiseitl

WATER MANAGEMENT IN MCLAREN TECHNOLOGY CENTER, LONDON, UK
英国伦敦麦克拉伦技术中心水管理项目
-1,2 ©Atelier Dreiseitl
-3 ©Nigel Young Foster + Partners
-4 © Foster + Partners
-5, 6,7,8 © Atelier Dreiseitl

Illustration Credits

MIXED-LIFESTYLE DEVELOPMENT,
CHANGCHUN VANKE, CHINA
中国长春万科综合生活区开发项目
-1,2,3,4,5,6,7 © Atelier Dreiseitl

GREEN ROOF,
CHICAGO CITY HALL, USA
美国芝加哥市政厅屋顶花园
-1 ©Cook & Jenshel, NG Creative
-2,3 ©Mark Farina
- 4 ©Atelier Dreiseitl

MITTELSTRASSE GEVELSBERG,
GERMANY
德国格勒斯伯格市中央街设计
-1,3,4,5,6 © Conne van D`Grachten
- 2 © Atelier Dreiseitl

MAYBACH CENTER OF EXCELLENCE,
STUTTGART, GERMANY
德国斯图加特迈巴赫卓越中心
-1,2,3,4,5,6,7 ©Atelier Dreiseitl

TANNER SPRINGS PARK,
PORTLAND, USA
美国波特兰纳斯普林斯公园
-1 ©J. Hoyer
-2 © T Good
-3 ©M Houck
-4 ©Atelier Dreiseitl
-5 ©Green Works
-6,7 ©Atelier Dreiseitl

SOLAR CITY, LINZ, AUSTRIA
奥地利林茨太阳城
-1 ©Pertlwieser/StPL
-2,3,4,5 ©Atelier Dreiseitl
-6,7 ©D Moet

QUEENS BOTANICAL GARDEN,
NEW YORK, USA
美国纽约皇后植物园
-1,2,3,6,7 ©ESTO
-4,5 ©Atelier Dreiseitl

ZHANGJIAWO
NEIGHBOURHOOD COMMUNITY,
TIANJIN
天津张家窝新镇社会山小区
-1 ©Shanghai Wisepool
 2,3 ©Atelier Dreiseitl
4,5,6 ©Shanghai Wisepool
7,8,9,10,11,12 ©Atelier Dreiseitl

SCHARNHAUSER PARK,
OSTFILDERN, GERMANY
德国奥斯特菲尔登沙恩豪瑟公园
-1,2,3,4,5,6,7 ©Atelier Dreiseitl

A CLIMATE ADAPTATION COMMUNITY,
ARKADIEN WINNENDEN, GERMANY
德国阿卡迪亚温嫩登气候适应型社区
-1 ©Doherty
-2,3,4,5,6,7,8,9 ©Atelier Dreiseitl
-10 ©Eble Architekten
-11,12,13 ©Atelier Dreiseitl

图片版权

GUORUI INTERNATIONAL INVESTMENT SQUARE, BEIJING, CHINA
中国北京国锐国际投资广场
-1 © Foster & Partners
-2,3,4,5,6,7,8 © Atelier Dreiseitl

CENTRAL FOUNTAIN, RITZ CARLTON, TIANJIN TAIANDAO, CHINA
中国天津泰安道丽思卡尔顿酒店中央水景
-1,2,3 ©Atelier Dreiseitl

WATER PLAYGROUND BUGA, KOBLENZ, GERMANY
德国科布伦茨BUGA嬉水游乐场
-1 ©Atelier Dreiseitl
-2 ©Doherty
-3 ©Atelier Dreiseitl

LIGHT AND WATER INSTALLATION, BREITEWEG
布莱特威戈"光与水"的装置
-1 ©Atelier Dreiseitl

TAIPEI PUBLIC ART PROJECT – WATER-ART RED SNAKE
台北公共艺术设计——红蛇水景艺术装置
-1,2,3 ©Atelier Dreiseitl

RED SWINGS, VENLO, NETHERLANDS
荷兰芬洛红色摇摆水景装置
-1,2,3 ©Atelier Dreiseitl

ART WALL, TANNER SPRINGS PARK, PORTLAND, USA
美国波特兰坦纳斯普林斯公园艺术墙
-1,2,3 ©Atelier Dreiseitl

WATER CURTAIN FOUNTAIN, HEINER-METZGER PLAZA, NEU-ULM, GERMANY
德国新乌尔姆,海纳–梅茨格广场,水幕喷泉
-1,2,3,4 ©Conne can D'Grachten

The Authors

Herbert Dreiseitl, born in 1955, is a sculptor, water artist and landscape architect. After his studies, he founded Atelier Dreiseitl in Überlingen on Lake Constance in 1980. He has realized numerous projects in the fields of stormwater management, water art and landscape architecture. He is active in lecturing worldwide and his work is widely published. He is guest professor at NUS Singapore and director at Ramboll LCL.

Dr. Eduard Kögel, born in 1960 in southern Germany. Studied Urban Design and Urban Planning in Germany and France. Working as critic, publisher, content manager and curator in the field of architecture and urban development. The focus of his work is the development in Asia and the Pacific Rim.

Prof. **Antje Stokman**, born in 1973, studied Landscape architecture at Leibniz University Hanover and at Edinburgh College of Art. From 2005~2010 she was assistant professor, focusing on Ecosystem design and watershed management' at Leibniz University Hanover. Since 2010 she is full professor and director of the Institute of Landscape Planning and Ecology at Stuttgart University.

Prof. **Wu Che**, born in 1955, is professor of environmental engineering at Beijing University of Civil Engineering and Architecture University and advisor committee member of China Green Building Council. He studied for a master's degree of water and wastewater engineering at Wuhan University of technology from 1980 to 1983. His major interest is urban stormwater management include urban flooding control and rainwater utilization, urban stormwater runoff pollution and CSO pollution control, ecological restoration of urban water environment and landscape planning.

Dieter Grau, born in 1963, gardener and accredited landscape architect, has been working for Atelier Dreiseitl since 1994. In 1996 he became head of the Landscape Architecture Department. He is Partner and manging director of AD and has led many urban projects around the world. He lectures and publishes frequently worldwide.

Gerhard Hauber, Landscape architect, partner and managing director of Atelier Dreiseitl, has been practicing landscape architecture and planning for more than 15 years. Since 1998 he has specialized in international project management. He is member of the German Sustainable Building Council (DGNB) and conducts training sessions for Auditors.

Khoo Teng Chye graduated in Civil Engineering from Monash University, Australia in 1975. He also holds a Master of Science in Construction Engineering and a Master of Business Administration from the National University of Singapore. Mr. Khoo is the Chief Executive and a Board Member of PUB, Singapore's national water agency since December 2003. He is Executive Director of the Environment and Water Industry Development Council.

Prof. **Wolfgang F. Geiger** is one of the few experts in natural water management. He is head of the Department of Water Management at the Universität Gesamthochschule in Essen, Faculty of Building.

作者简介

赫伯特·德赖赛特尔，生于1955年，雕塑家、水景艺术家和景观设计师。于1980年在康斯坦茨湖滨的于波林根成立德国戴水道设计公司。他实践完成了雨水管理、水景艺术和景观设计等领域的众多项目。多年来，他活跃在世界各地进行讲学，作品广泛流传。如今他作为新加坡国立大学客座教授并担任丹麦安博国际工程咨询公司"宜居城市研究所"主任。

爱德华·科格尔博士，1960年出生于德国南部。在德国和法国学习、研究城市设计与城市规划。目前，他作为建筑和城市发展领域的评论家、出版人、内容管理师和策展人，工作重点在亚洲和环太平洋地区的发展方面。

安茱·斯托克曼教授，生于1973年，就读于莱布尼兹汉诺威大学和爱丁堡艺术学院学习景观设计。2005年~2010年间在莱布尼兹汉诺威大学任职助理教授，专注于生态系统设计和流域管理。自2010年以来，作为斯图加特大学教授和景观规划与生态研究所主任。

车伍，生于1955年，毕业于武汉工业大学给水排水工程，硕士学位。现为北京建筑大学环境工程系教授，中国城市科学研究会节能与绿色建筑专业指导委员会委员。主要研究方向为城市雨洪控制利用，城市径流的非点源污染及CSO污染控制，城市洪涝控制和城市水环境生态修复与水景观方面的研究、规划设计与工程实施等。

迪特尔·格劳，生于1963年，园艺家、注册景观设计师，1994年进入德国戴水道设计公司至今。1996年成为景观设计部门负责人，目前作为公司合作人和运营总监，参与世界各地的项目设计主持，并在世界各地进行演讲。

格哈德·豪博，景观设计师，德国戴水道设计公司合伙人，执业景观规划设计行业超过15年。自1998年以来，一直专注于国际项目的管理。作为德国可持续建造委员会（DGNB）的成员之一，负责审核者的资格培训。

邱登才，1975年毕业于澳大利亚莫纳什大学土木工程专业，并拥有新加坡国立大学施工工程理学硕士学位以及工商管理硕士学位。自2003年12月起，邱先生作为新加坡公用事业局和新加坡国家水务局行政官员和董事会成员，同时他也作为环境和水工业发展委员会执行董事。

沃夫冈·F·盖格教授是为数不多的天然水资源管理专家，并作为杜伊斯堡–埃森大学建造学院水资源管理系主任。

Atelier Dreiseitl

Partners
Herbert Dreiseitl
Dieter Grau
Gerhard Hauber

Directors
Tobias Baur
Claudia Bross
Stefan Brueckmann
Christoph Hald
Rudolf Mager
Leonard Ng
Hendrik Porst
Zheng Sun

Associates
Jessica Read
Alexander Rohe
Pedro Santa-Rivera
Vera Sieber
Hao Wu
Nengshi Zheng
Florian Zimmermann

AD Germany Überlingen Team
Jeremy Anterola
Claudia Bross
Stefan Brueckmann
Herbert Dreiseitl
Julia Dreiseitl
Berthold Flieger
Dieter Grau
Christoph Hald
Gerhard Hauber
Mariusz Hermansdorfer
Chaojun Li
Rudolf Mager
Isabell Maser
Nora Menzel
Constantin Möller
Ursula Münzberg
Tom Patterson
Hendrik Porst
Jessica Read
Alexander Rohe
Bernd Schernau
Ulrike Schmidt-Henning
Vera Sieber
Falko Stengel
Sebastian Walker
Nengshi Zheng
Christof Würtle
Gustavo Gläser
Andreas Bockemühl

AD Singapore Team
Tobias Baur
Nurilsa Binte Taip
Agnes Chain
Jade Chin
Wi Ming Chin
Jay Coro
Phibang Do
Kenya Endo
Sam Yik Hong
Madeline Leong
Sheetal Muralidhara
Leonard Ng
Pedro Santa-Rivera
Ryan Shubin
Anton Siura
Browyn Tan
Christina Ting
Chris Yip
Melissa Yip
Chanida Suebpanich

AD China Beijing Team
Zheng Sun 孙峥
Florian Zimmerman 付德景
Hao Wu 吴昊
Ulrich Hoffmann
Philipp Nedomlel
Matthias Erlen
Feng Ouyang 欧阳凤
Vivian Ouyang 欧阳红
Bing Cao 曹冰
Zhirong Yan 闫志荣
Jincao Li 李劲草
Pei Dang 党培
Yinshi Jin 金银实
Shengnan Tao 陶胜男
Peng Zhou 周鹏
Li Zhang 张黎
Cathy Lv 吕焕来
Yuefeng Pan 潘越丰
Qiong Wu 吴琼
Ting Shi 石婷
Feng Gao 高枫
Xuluan Chen 陈旭娈
Sai Chen 陈赛
Xinyi Zhang 张新伊
Zhenrong Zhu 朱珍蓉
Junying Yuan 袁军营
Ran Jia 贾然
Mengwei Li 李孟伟

Waterscapes Innovation.
水敏性创新设计